2110

STATISTIC

For Business, Finance, a.

To
VANESSA

STATISTICS

For Business, Finance, and Accounting

J. P. DICKINSON
M.A., M.Sc., Ph.D, A.M.B.I.M.

Senior Lecturer in Commerce
University of Western Australia

MACDONALD AND EVANS

MACDONALD & EVANS LTD.
Estover, Plymouth PL6 7PZ

First published April 1976
©
MACDONALD AND EVANS LIMITED
1976

ISBN: 0 7121 1939 6 (Hardback)
ISBN: 0 7121 1948 5 (Paperback)

Printed in Great Britain by
Richard Clay (The Chaucer Press) Ltd
Bungay, Suffolk

PREFACE

The years since 1960 have witnessed an enormous upsurge in the application of statistical techniques to an increasingly complex business world. Statistical analyses bring both objectivity and efficiency to future planning and decision-making, and to the interpretation of vast amounts of numerical data. Accordingly, in order to service the needs of the modern businessman and business student, a number of excellent "business statistics" texts have appeared over the last few years.

The author has himself been concerned for several years with teaching statistics to general management and business students, and—more particularly—to students of accounting and finance. In his experience the latter students have been at a great disadvantage, since none of the available texts is really suitable for their needs. Indeed, it has been said that statistics is a subject little understood by accountants, and it may well be that the reason for this lies in the lack of a text which approaches statistics from a financial viewpoint, and demonstrates the relevance of the techniques in the financial world.

The present volume has been written in an attempt to fill this gap. It is based on material which has been developed in the classroom, and through discussions with practising accountants, with students working for professional examinations, and with teachers of accounting. The book assumes a knowledge of mathematics equivalent to Ordinary Level G.C.E., although certain parts of the syllabus, such as trigonometry, are not required. Some work on sets, however, is necessary and has been included since the topic is not yet incorporated in all O-level courses. In its coverage the book is intended to meet the examination requirements of professional bodies such as the Association of Certified and Corporate Accountants, the Institutes of Chartered Accountants, and the Institute of Cost and Management Accountants. Bearing in mind the growing tendency towards a graduate profession, the text should also be very relevant to the university or polytechnic student working for a degree or diploma with an accounting component. It is also hoped that the readership will include the professional accountant and businessman wishing to acquaint himself with the current applications of statistics.

Without using unnecessary rigour, the author has sought to present

the material in a modern and hopefully interesting manner, in order
that the reader can understand not only what he is doing at each
stage, but why he is doing it. To assist in the learning process a wide
range of exercises has been included, together with a large number of
worked examples. Suggestions for further reading are provided in the
Bibliography.

ACKNOWLEDGMENTS

I should like to express my gratitude to colleagues in the Department of Accounting and Finance at the University of Lancaster, who have read through earlier drafts of this book and furnished many helpful comments.

My sincere thanks are also due to Mrs. Valerie Wilson, for her invaluable assistance in the typing of an almost illegible manuscript.

To my wife, Christine, however, I owe the greatest debt. She has not only been responsible for the typing (and retyping) of a considerable part of this book, but has provided encouragement at all stages of its preparation, and an abundance of patience and understanding in my many periods of silent contemplation! Without her help the book would not have been completed.

J.P.D.

Lancaster, November 1975.

ACKNOWLEDGMENTS

CONTENTS

PAGE

PREFACE   v

ACKNOWLEDGMENTS   vii

CHAPTER

1. Frequency Distributions   1
   Introduction   1
   The graphical summary of data   1
   Summary statistics   13

2. Probability   26
   Introduction   26
   Sets   26
   Experiments and events   34
   Evaluation of probabilities   38
   Relations between events   46

3. Joint and Conditional Probabilities   50
   Introduction   50
   Joint probability   50
   Conditional probability   54
   Permutations and combinations   59

4. Properties of Probability Distributions   68
   Introduction   68
   Mean and variance of probability distributions   78
   Some general comments on distributions   85
   More than one random variable   88

5. Discrete Probability Models   97
   Introduction   97
   Some special discrete probability models   97

6. Continuous Probability Models   116
   Introduction   116
   The exponential distribution   116
   Beta distribution   119
   The normal distribution   121

7. Sampling Theory and Estimation   137
   Introduction   137
   Sampling schemes   138
   Sampling objectives   141
   Estimation   142

CHAPTER	PAGE
8. Statistical Hypothesis Testing	165
Introduction	165
Tests concerning population means	165
Two sample tests involving population means	173
Tests concerning proportions	181
Finite population correction	186
Errors in statistical hypothesis testing	186
9. Contingency Tables and Analysis of Variance	195
Introduction	195
Contingency tables	195
Analysis of variance	206
10. Non-parametric Statistics	219
Introduction and definition	219
Tests involving one sample	219
Tests involving two matched samples	228
Tests involving two independent samples	233
11. Regression and Correlation Analysis	236
Introduction	236
Simple linear regression	236
Correlation	248
Extensions to simple linear regression analysis and correlation	255
12. Time Series Analysis and Index Numbers	262
Introduction	262
Long-term trend	263
The seasonal factor	270
The cyclical factor	275
Index numbers	276
13. The Elements of Statistical Decision Theory	283
Decisions and decision criteria	283
Bayesian decision analysis	287

APPENDIXES

I. Bibliography	297
II. Table A: Areas of the standardised normal probability distribution	300
III. Table B: Random sampling numbers	301
IV. Table C: t distribution	302
V. Table D: Chi-square distribution	303
VI. Table E: F-distribution—upper 5% points	304
VII. Table E: F-distribution—upper 1% points	305
INDEX	307

LIST OF ILLUSTRATIONS

FIG.   PAGE

1. Histogram of distribution of incomes   2
2. Incorrect histogram of data of Table 2   4
3. Correct histogram of data of Table 2   5
4. Histogram of sales revenue/week   8
5. Relative frequency histogram of sales revenue/week   8
6. Superposition of absolute frequency histograms   9
7. Superposition of relative frequency histograms   11
8. Frequency polygon derived from Fig. 1   11
9. Cumulative frequency distribution of incomes   13
10. Ogive for data of Table 1   14
11. Relations between sets   29
12. Unions and intersections   31
13. Venn diagram   33
14. Venn diagram—set of 27 possible behaviours   35
15. Typical pattern of result of throwing die   38
16. Two events concerning three-share portfolio   41
17. Events N and K 42
18. Events L and J 43
19. General relationship between any three sets A, B, C 44
20. Complementary events   47
21. Uniform continuous probability distribution   73
22. General continuous probability density function   74
23. Exponential distribution   75
24. Relative frequency distribution of asset lives   77
25. Theoretical probability distribution of asset lives   77
26. Graph of probability density function   80
27. Symmetrical probability distributions   85
28. Modality of distributions   86
29. Skewness   86
30. Relationship between measures of centrality   87
31. Distributions with equal means and variances but oppositely skewed   87
32. Probability distribution function   88
33. Graphical representation of joint probability density function   91
34. The binomial distribution   102
35. The geometric distribution   106
36. Poisson distribution for different values of λt 108
37. Comparison of binomial and Poisson distributions   113
38. Exponential distribution   117
39. Beta distribution   119
40. Histogram of experimental results   121
41. Probability density function of $N(0, 1)$ 122
42. Probability distribution function, $\Phi(x)$, of $N(0, 1)$ 124

FIG.   PAGE

43. Distribution function of $N(0, 1)$ for negative values of x 124
44. Probabilities associated with a standard normal variable   125
45. Distributions of normal random variables   128
46. Daily takings in millinery store   129
47. Sampling schemes   140
48. Variation in sample mean   144
49. Effect of increased sample size on distribution of sample mean   145
50. Confidence intervals for μ 147
51. Comparison of standard normal and Student's t-distribution   152
52. Acceptance/Rejection regions   167
53. Test of H_0: $\mu = 30$ against H_1: $\mu > 30$ 168
54. Test of H_0: $\mu = 8.5$ against H_1: $\mu < 8.5$ 169
55. Test of H_0: $\mu \geqslant 30$ against H_1: $\mu < 30$ 170
56. Test of H_0: $\mu = 100$ against H_1: $\mu < 100$ 172
57. Test of H_0: $\mu_1 \geqslant \mu_2$ against H_1: $\mu_1 < \mu_2$ (small sample)   177
58. Test of H_0: $\mu_2 - \mu_1 = 3.20$ against H_1: $\mu_2 - \mu_1 > 3.20$ 178
59. Type II error in test H_0: $\mu = 30$ against H_1: $\mu \neq 30$, when in fact $\mu = 32$ 188
60. Operating characteristic curve   188
61. Power curve   188
62. Type II error of test H_0: $\mu = 30$, H_1: $\mu \neq 30$ for true values of μ of 30.1 and 34   189
63. Type II error of test H_0: $\mu = 30$, H_1: $\mu > 30$ 189
64. Operating characteristic curves for upper-, lower- and two-tailed tests   191
65. Effect on a of decreasing β and on a and β of increasing n 192
66. Control of Types I and II errors—choice of n 193
67. χ^2-distribution for various degrees of freedom   199
68. χ^2-distribution—four degrees of freedom   200
69. General form of F-distribution   209
70. Distribution of $F_{2, 15}$ 209
71. Possible non-linear relationships between x and y 237
72. Scatter diagram of observed pairs (x, y) 237
73. Dangers of extrapolation   240
74. True and estimated least squares regression line   241
75. Scatter diagrams—perfect correlation, imperfect correlation, no correlation   248
76. Analysis of variation in y-values   249
77. Different values of $\Sigma(x - \bar{x})(y - \bar{y})$ 251
78. Fitting of several least squares lines segments   256
79. Fitting of least squares curve $y = Ae^{mx}$;   256
80. Fitting of least squares curve $y = A/x + B$ 257
81. Scatter diagram—wage-rate/no. of defectives   258
82. Time series of (hypothetical) tour operator's income   263
83. Cost graph, showing least squares trend line, extrapolation of linear trend and deseasonalised series   265
84. Typical trend lines   269
85. Ratio to moving average   273
86. Total probability theorem   288

LIST OF TABLES

TABLE   PAGE

1. Distribution of incomes in a particular district (equal class intervals)   2
2. Distribution of incomes in a particular district (open-ended; unequal class intervals)   3
3. Expanded version of data of Table 2   5
4. Sales revenue over 52-week period   7
5. Sales revenue over 100-week period   9
6. Distribution of income (second region)   10
7. Cumulative frequency table for income data of Table 1   12
8. Annual profits 1963–72   18
9. Distribution of daily takings   18
10. Price of share over 200 days   19
11. Coded share prices from Table 10   20
12. Budget variances   21
13. Daily takings   24
14. Classification of 1000 shares (dependence between factors)   50
15. Classification of 1000 shares (independent factors)   51
16. Distribution of revenue from new product   54
17. Probability distribution of service lives of two machines   56
18. Permutations of three projects from five   63
19. Distribution of asset lives   76
20. Joint probability table of work-in-process and finished goods   89
21. Probability table of random variable $(X + Y)$ 91
22. Probability table of random variable $(X + Y)^2$ 93
23. Tax information on ten individuals   96
24. Use of normal approximation to binomial distribution   134
25. Random sample of invoices   146
26. Department sales—budgeted and actual   179
27. Summary of hypothesis testing   187
28. Breakdown of accounts of Natclay Bank Ltd. (particular branch)   196
29. Breakdown of sample of Natclay Bank Ltd. head office accounts   196
30. Expected frequencies in Natclay sample   197
31. Observed breakdown of accounts at four Natclay branches   201
32. Expected breakdown of accounts at four Natclay branches   201
33. Breakdown of incomes of Ardour University graduates   202
34. Expected breakdown of incomes of Ardour University graduates   203
35. Maintenance costs per year (£)   206
36. Modified maintenance costs—eliminating within groups variation (conceptual)   207
37. Analysis of variance table for maintenance costs   210
38. ANOVA table for typists' coffee break times   212
39. Maintenance costs: by machine types and by engineers   214
40. Two-way ANOVA table of maintenance costs   216
41. Coded data for salesman   217

TABLE   PAGE

42. ANOVA table for data on salesmen   217
43. Analysis of demand for car—goodness-of-fit test for Poisson distribution   224
44. Analysis of costs of chemical—goodness-of-fit test for normal distribution   226
45. Staff scores for "Samplan" and "Testy"   228
46. Wilcoxon Signed Rank-Sum test using machine output data   230
47. Rank correlation between managers   232
48. Output of jaggers and total cost for Magwitch Ltd.   239
49. Costs of borborofluoride   264
50. Five-point moving averages using data from Table 49   266
51. Six-point centred moving average using data from Table 49   267
52. Regression analysis of data from Table 49 using coded times over 60-month period   268
53. Calculation of ratio to centred 12-point moving average   271
54. Complete listing of ratio to moving average (as calculated in Table 53)   272
55. Seasonal indices (calculated as the median of the four values of ratio to moving average given in Table 54)   272
56. Calculation of cyclical index   276
57. Simple price relatives of borborofluoride (May 1974 = 100)   277
58. (Unweighted) aggregative price index (1971 = 100)   277
59. (Unweighted) aggregative price index (1971 = 100)—effect of measuring price of milk in gallons rather than pints   278
60. Calculation of Laspeyres and Paasche indexes   279
61. Average price relatives (1971 = 100)   280
62. Change of base-period   281
63. Pay-off table for investment decision   284
64. Opportunity loss table for investment decision   285
65. Conditional probabilities associated with market research organisation's report   293
66. Posterior event probabilities   294
67. Posterior expected pay-offs   294

CHAPTER 1

FREQUENCY DISTRIBUTIONS

INTRODUCTION

One purpose of statistical analysis is to reduce a mass of data into a more compact form which highlights general trends, relationships between variables such as costs and revenues, patterns of behaviour of profit figures or of cash flows, and so on. It is by no means always the case that the data is in numerical form, but even when it is not, the objective is still to provide a quantitative (and, therefore, hopefully objective) way of distilling the essential features from the data.

A statistic is simply a numerical quantity derived from a body of data. The least, the greatest, the average of a set of numerical quantities are all perfectly legitimate statistics. It is true to say that many of the statistics defined in this broad sense are not particularly useful. On the other hand one can equally say that no one statistic is satisfactory for all purposes. The skill of the statistician lies in the selection of the most appropriate statistics and form of analysis in a particular situation, and in the interpretation of this analysis.

In this chapter we shall chiefly be concerned with a type of statistic loosely defined as *descriptive*. Since such a statistic provides in one way or another a summary of a body of data, it is appropriate also to discuss the graphical presentation of data.

THE GRAPHICAL SUMMARY OF DATA

1. The simple histogram. One of the simplest and most useful means of condensing numerical information graphically is through the *histogram*. Table 1 gives the annual incomes to the nearest pound in a particular district of the 15,000 earning members of the community.

The data have been collected in the course of a national survey, and are to be used in planning future development in the area. For example, general retailers, supermarkets, and banks would find this sort of information of great value in determining expansion policies.

Notice that the raw data—comprising the 15,000 exact incomes—have already been simplified by placing incomes into categories £250 "wide." This enables a considerable condensation of the original data, although admittedly the fine detail has been lost. The point is, of course, that only the broad picture (so long as it is a fair one) is of interest to the planners.

Table 1. Distribution of incomes in a particular district
(equal class intervals)

Income, I, in £ per annum	Number of persons
$1000 \leqslant I < 1250$	600
$1250 \leqslant I < 1500$	800
$1500 \leqslant I < 1750$	1200
$1750 \leqslant I < 2000$	3000
$2000 \leqslant I < 2250$	3200
$2250 \leqslant I < 2500$	2700
$2500 \leqslant I < 2750$	1800
$2750 \leqslant I < 3000$	900
$3000 \leqslant I < 3250$	500
$3250 \leqslant I < 3500$	300

Table 1 is an example of a *frequency table*, and its contents give a *frequency distribution*. The frequency in a particular category is known as the *class frequency*. This tabular form lends itself very conveniently to the graphical presentation referred to as a histogram (Fig. 1).

The clear advantage of such a display is that it gives a visual impression of the distribution at a glance.

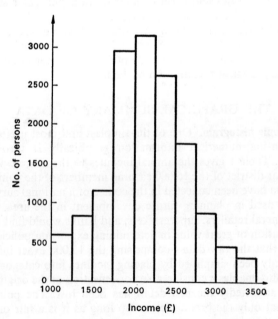

FIG. 1.—Histogram of distribution of incomes.

Notice that, following normal practice, the blocks have been drawn abutting, although strictly speaking none of the data values lie within the range £1249–£1249½, for example. This is of little importance, since the object of the histogram is to provide simplicity and clarity. Really on the same point, the class intervals are taken as having a width of £250 (being the difference between the lower ends of two adjacent classes) rather than, for instance, of £1249½ − £1000 = £249½. Because of the manner in which the data have been categorised in the table the preparation of the histogram is quite straightforward.

The class intervals are disjoint (*i.e.*, do not overlap) so that there is no ambiguity concerning the numbers of persons in any range, and none concerning the heights of rectangles in the histogram. This would not have been the case if the intervals had been given as $1000 \leqslant I \leqslant 1250$, $1250 \leqslant I \leqslant 1500$, etc.

2. Further remarks on histograms. Table 1 was, perhaps, rather artificial, in that the data would not normally be "cut off" so clearly at each end. Suppose instead that we had been given the data in Table 2

Table 2. Distribution of incomes in a particular district (open-ended; unequal class intervals)

Income, I, in £ per annum	Number of persons
$I < 1250$	600
$1250 \leqslant I < 1500$	800
$1500 \leqslant I < 1750$	1200
$1750 \leqslant I < 2000$	1300
$2000 \leqslant I < 2250$	3200
$2250 \leqslant I < 2500$	2700
$2500 \leqslant I < 2750$	1800
$2750 \leqslant I < 3000$	900
$3000 \leqslant I < 3250$	500
$3250 \leqslant I$	300

and wished to present it in histogram form. The point here is that, whereas in theory incomes go down to zero and up to say £50,000 or even £100,000, in the extreme ends of the frequency tables there may be several class intervals with only a few (or sometimes no) members. Immediately this raises the problem of adequate representation on the histogram. Normal practice is to take reasonable values for the extremes—in this case £500 and £4500. One could then construct rectangles of heights 600 and 300 on bases £500–£1250 and £3250–£4500 as in Fig. 2.

This histogram gives, however, a quite misleading picture of the

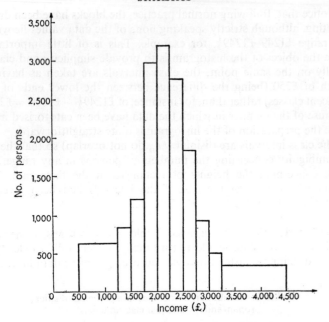

FIG. 2.—Incorrect histogram of data of Table 2.

distribution of incomes since the "tails" are too fat. To correct this, one need only realise that the 600 persons with incomes of less than £1250 have been assumed to be distributed over a range £500–£1250. In the absence of detailed information one can do no better than to assume that they are evenly distributed over this range (*i.e.*, 200 in each of the three intervals of £250). Similarly, at the other extreme, there are five intervals of £250 in the range £3250–£4500. It is reasonable to assume that each of these "sub-ranges" contains 60 persons. Accordingly, assuming that we have only the data given in Table 2, we may present this in the revised form of Table 3, which leads to the modified histogram of Fig. 3.

Of course the general method suggested here of dealing with open-ended extreme classes in the frequency tables does not lead to an unique histogram. Even so, the method does give a much fairer impression of the distribution than Fig. 2, whatever extreme values are assumed.

A point to be emphasised is that in a histogram it is not the heights of the rectangles that represent class frequencies, but their areas. In Table 1 the class intervals were of equal length (*viz.* £250). The bases of all the rectangles were therefore equal, and their areas were in the same proportions as their heights. Here then, the rectangles were

Table 3. Expanded version of the data of Table 2

Income, I, in £ per annum	Number of persons
$500 \leqslant I < 750$	200
$750 \leqslant I < 1000$	200
$1000 \leqslant I < 1250$	200
$1250 \leqslant I < 1500$	800
$1500 \leqslant I < 1750$	1200
$1750 \leqslant I < 2000$	3000
$2000 \leqslant I < 2250$	3200
$2250 \leqslant I < 2500$	2700
$2500 \leqslant I < 2750$	1800
$2750 \leqslant I < 3000$	900
$3000 \leqslant I < 3250$	500
$3250 \leqslant I < 3500$	60
$3500 \leqslant I < 3750$	60
$3750 \leqslant I < 4000$	60
$4000 \leqslant I < 4250$	60
$4250 \leqslant I < 4500$	60

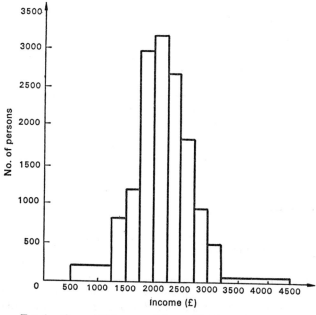

FIG. 3.—Correct histogram of data of Table 2 (via Table 3).

legitimately drawn with heights in the same proportions as the class frequencies. In Table 2, extreme class intervals were proposed whose lengths were different from those in the rest of the table. The method of dealing with this can be considered in two ways. First, since the lowest class interval was three times the "standard" length of £250, in order to maintain the proportional relationship between areas and frequencies, the rectangle is drawn one-third of the height one might first expect. Alternatively, one can break down the lowest class interval into three sub-intervals, each of the "standard" length of £250, and evenly distribute the 600 persons over these sub-intervals. The rectangles on each of these sub-intervals are then drawn with their heights (and therefore their areas) in proportion to the numbers of persons in them.

A final comment concerns the vertical scale on a histogram. In the preceding tables it may seem misleading—in view of the assertion that areas represent frequencies—to label the ordinate "number of persons." However, by dividing the larger intervals into smaller intervals of "standard" length, and correcting the heights of the rectangles accordingly, account has been taken of the relationship between areas and frequencies.

As a result, there is no harm in labelling the ordinate "number of persons."

Exercises on Section 2

1. A large manufacturing company requiring improvement of its pension scheme has provided an insurance company with the following table of the age distribution of its employees:

Age (A) in years	No. of employees
A < 20	500
20 ≤ A < 30	2000
30 ≤ A < 35	2200
35 ≤ A < 40	2800
40 ≤ A < 55	2100
A ≥ 55	600

Present this information in the form of a histogram, clearly stating any assumptions that you make.

2. The average variable costs/unit are calculated on a weekly basis by dividing the total variable costs in a week by the number of units produced. Data is collected over a period of one year, and presented in the following frequency table:

Average variable costs/unit (C) (P)	Number of weeks
$0 \leqslant C < 3$	1
$3 \leqslant C < 5$	3
$5 \leqslant C < 6$	4
$6 \leqslant C < 7$	5
$7 \leqslant C < 8$	10
$8 \leqslant C < 9$	11
$9 \leqslant C < 10$	9
$10 \leqslant C < 11$	8
$11 \leqslant C < 13$	1
Total	52

Draw a histogram to present this data.

3. Are frequency tables in which class intervals are given in the form
   (a) 0–100, 100–200, etc.
   (b) 0–99, 100–199, 200–299, etc.
suitable for preparing histograms? Explain.

3. Relative frequency histograms. Table 4 gives the sales revenue per week from a certain product over a 52-week period.

Table 4. Sales revenue over 52-week period

(1) Sales Revenue (S) in £000	(2) Number of weeks	(3) Number of weeks as a percentage of total
$1 \leqslant S < 2$	4	7·69
$2 \leqslant S < 3$	6	11·54
$3 \leqslant S < 4$	9	17·31
$4 \leqslant S < 5$	15	28·84
$5 \leqslant S < 6$	10	19·23
$6 \leqslant S < 7$	6	11·54
$7 \leqslant S < 8$	2	3·84
Total	52	100

The histogram based on columns (1) and (2) is easily constructed (Fig. 4). Alternatively, using column (3), which gives the relative frequency of each category of sales revenue in terms of the percentage number of weeks it occurs, a *relative frequency histogram* can be produced (Fig. 5).

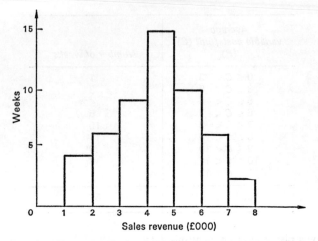

FIG. 4.—Histogram of sales revenue/week over a 52-week period.

The profile of this second histogram is exactly the same as that of the original. However, the essential difference between the two is that the area underneath the second is unity (or 100%).

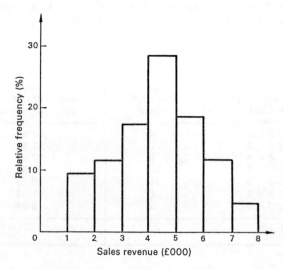

FIG. 5.—Relative frequency histogram of sales revenue/week.

An advantage of relative frequency presentation can be illustrated by considering additional data on a further product collected over a 100-week period (Table 5).

Table 5. Sales revenue over 100-week period

(1) Sales Revenue (S) in £000s	(2) Number of weeks	(3) Number of weeks as a percentage of total
$1 \leqslant S < 2$	7	7·00
$2 \leqslant S < 3$	15	15·00
$3 \leqslant S < 4$	15	15·00
$4 \leqslant S < 5$	33	33·00
$5 \leqslant S < 6$	17	17·00
$6 \leqslant S < 7$	9	9·00
$7 \leqslant S < 8$	4	4·00
Total	100	100

Superposition of the *absolute* frequency histogram for the second product on that for the first leads to Fig. 6.

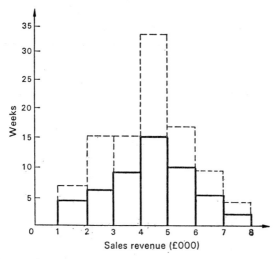

Fig. 6.—Superposition of (absolute frequency) histograms for sales revenue from two products.

Because of the differences in total frequencies (*i.e.*, 52 and 100) it is difficult to compare the two patterns of sales revenues using this combined absolute frequency histogram. Superposition of the *relative* frequency histograms enables this comparison to be made much more readily (Fig. 7 on p. 11).

Exercises on Section 3

4. Construct two relative frequency histograms to compare the income information of Table 1 with the information on a second region given in Table 6.

Table 6. Distribution of income (second region)

Income, I, in £ per annum	Number of persons
I < 2000	100
2000 ≤ I < 2500	250
2500 ≤ I < 3000	400
3000 ≤ I < 3250	1300
3250 ≤ I < 3500	2600
3500 ≤ I < 3750	4000
3750 ≤ I < 4000	3100
4000 ≤ I < 4250	2200
4250 ≤ I < 4750	1270
4750 ≤ I < 5250	500
I ≥ 5250	220

5. The performances of candidates in a certain examination of one of the professional accounting bodies in 1973 and 1974 are summarised below:

Mark obtained (%)	Number of candidates	
	1973	1974
15	43	62
15–24	65	86
25–34	115	160
35–39	147	208
40–44	91	142
45–49	108	161
50–54	135	190
55–59	129	211
60–64	111	157
65–69	87	145
70–79	102	148
80	36	50
Total	1169	1720

Prepare two relative frequency histograms to compare the patterns of results.

4. Frequency polygons. The frequency polygon is a very simple extension of the histogram. Returning to Fig. 1, the midpoints of the tops

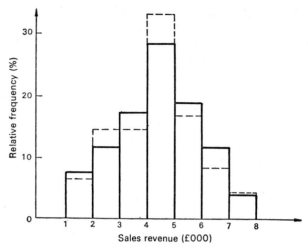

FIG. 7.—Superposition of relative frequency histograms for sales revenue from two products.

of each rectangle (whether the rectangles are, as in this case, on equal bases or not) are joined together in order by straight lines (Fig. 8). The beginning and end points on the horizontal axis are taken respectively as being half a class interval (in this case £125) below the

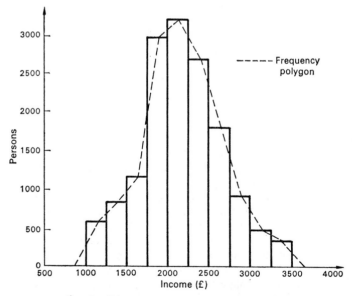

FIG. 8.—Frequency polygon derived from Fig. 1.

lowest and above the highest classes. After removing the histogram blocks the remaining figure is the *frequency polygon*. Its only real advantage is that it dispenses with the artificial stepped appearance of the parent histogram, providing a more realistically continuous appearance to the data.

5. Cumulative frequency distributions and ogives. An alternative and sometimes more useful way of graphically displaying tabulated data is by constructing a cumulative frequency distribution. With reference again to Table 1, the frequency table is first modified slightly in an obvious way to give a *cumulative frequency table* (Table 7). The

Table 7. Cumulative frequency table for income data of Table 1

Income, I, in £ per annum	Number of persons
$I < 1250$	600
$I < 1500$	1400
$I < 1750$	2600
$I < 2000$	5600
$I < 2250$	8800
$I < 2500$	11,500
$I < 2750$	13,300
$I < 3000$	14,200
$I < 3250$	14,700
$I < 3500$	15,000

cumulative frequency distribution can be constructed in an analogous way to the histogram, yielding Fig. 9.

Each block must be at least as high as its left-hand neighbour, since the cumulative frequency can never decrease with increasing income. In general the heights of the blocks increase from left to right, and two adjacent blocks will be of identical height only if the corresponding class interval has no members. Care must be taken to label the vertical axis as "cumulative frequency" in order to avoid misinterpretation of the graph as a normal histogram.

A common method of displaying cumulative frequencies is in the form of an *ogive*. Using the same data, points may be plotted on a graph with the cumulative frequency on the vertical axis, and the upper boundary of a class interval on the horizontal axis (Fig. 10).

As a matter of convenience the curve has been extrapolated to pass through the origin, since there are no persons with an income less than zero. The elongated S-shape of the ogive obtained here is characteristic of a parent histogram which tails off on either side of a single peak.

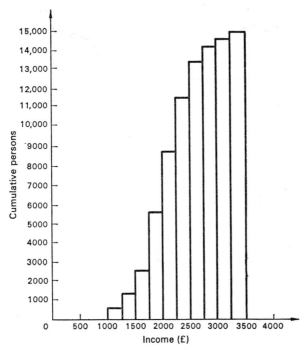

Fig. 9.—Cumulative frequency distribution of incomes.

Exercises on Section 5

6. Construct the cumulative frequency distributions and ogives for the following:
   (*a*) The data of Question 1.
   (*b*) The data of Question 2.
   (*c*) The data of Tables 4 and 5, on the same graph and in terms of cumulative relative frequencies.
   (*d*) The data of Question 5 on the same graph and in terms of cumulative relative frequencies.

SUMMARY STATISTICS

6. Measures of central tendency. The reader will be aware that a vast amount of numerical information is often available concerning the financial operations of a firm. For a particular purpose it is usually necessary to condense this information into a more manageable form. A department store, for instance, might collect information on each line that it sells over daily, weekly, and monthly periods. This minute detail is of undoubted use to a department head in determining his

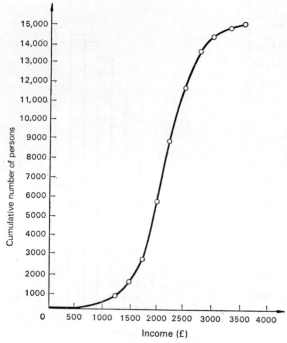

FIG. 10.—Ogive for data of Table 1.

reordering policy. To the top management of the store, however, who are concerned with overall profitability and with future development policy the fact that 30 combs were sold last Thursday is of no interest. Of more use to them are summary figures indicating trends and overall performance.

Suppose that the raw data consists of the daily takings of a particular department in the store over a period of 500 shopping days. Two features of a distribution of numerical values are especially useful for summary purposes—a *measure of central tendency* (indicating, rather loosely, the "middle" of the distribution), and a *measure of dispersion* about the centre. Taking the first of these, there are four methods of measuring centrality in common usage—the arithmetic mean, the median, the mode, and the geometric mean. In the development of statistical techniques the arithmetic mean is the most important of these, and will be used almost exclusively in this book. However, for completeness, the definitions of all four are given below, and explained in the context of the 500 daily takings.

The *arithmetic mean* of a set of measurements is the sum of the measurements divided by their number.

To calculate the mean takings, the total of the 500 figures is found, and is divided by 500.

An important property possessed by the mean is the ease with which it can be calculated and expressed mathematically as a simple combination of the n actual data figures $x_1, x_2, \ldots, x_n$. Thus,

$$\text{mean} = \sum_{i=1}^{n} x_i/n$$

(The summation notation $\sum_{i=1}^{n} x_i$ (Σ = Greek "sigma") is a concise way of writing $(x_1 + \ldots + x_n)$.)

The *mode* is the most frequently occurring value in a set of measurements.

If £830 occurs most often amongst the 500 daily figures, then this is the modal value. It may be that two (or more) figures occur equally often, and yet more often than any others. In this case the distribution has two (or more) modes.

Although the significance of the mode is readily understood, the fact that it may not be unique can be a disadvantage.

Furthermore the mode cannot be expressed in simple algebraic terms. This has resulted in few statistical procedures being developed which depend on the mode.

If a set of measurements is listed in ascending order of magnitude, the *median* is the middle measurement if the set contains an odd number of members, and the arithmetic average of the middle two if it contains an even number of members.

If there are 249 takings of lower value than £820, and 249 of higher value than £825, the median is

$$\frac{£820 + £825}{2} = £822\cdot50.$$

The median has the disadvantage that it involves the laborious arrangement of the data in order of magnitude. Additionally it has the same defect as the mean, in that it cannot be expressed simply in a mathematical form.

NOTE: By definition the median divides the data into two halves: one containing values greater than the median, and the other containing values less than the median. In a natural extension of this definition a distribution may sometimes be divided into four by *quartiles*, into tenths by *deciles*, and into hundredths by *percentiles*.

A property possessed by both the median and mode as compared with the mean is that of insensitivity to the occasional very high or very low measurement. This is easily illustrated by consideration of the following two sets of figures:

(a) 0, 1, 1, 1, 3	(b) 0, 1, 1, 1, 400
Mean = 1·2	Mean = 80·6
Mode = 1·0	Mode = 1·0
Median = 1·0	Median = 1·0

Here the two sets of figures differ markedly, but this is not reflected in the modal and median values. From this argument the mean can be relied upon to provide a more sensitive measure of centrality than the median or mode. On the other hand, this very sensitivity could be argued to be disadvantageous since the mean is influenced by the occasional unusually high or low datum value.

If the set of measurements does not contain any negative or zero values, the *geometric mean* is calculated by finding the nth root of the product of all n measurements. In our example we would have to multiply all 500 daily takings together, and then take the 500th root (*i.e.*, the number which when multiplied by itself 500 times would give the same total). Obviously the geometric mean is laborious to calculate. Furthermore, it is a meaningless statistic when any of the measurements are zero or negative. Its main value is in growth situations— such as a series of annual percentage increases—where an average rate of growth is required. In such cases it is the only measure of centrality which is appropriate.

Worked Examples

1. Find the mean, median, and mode of the following distribution of children per family:

No. of families	No. of children per family
6	0
11	1
20	2
15	3
8	4
4	5
2	6

Answer

The total number of families is 66. If we were to list all the families along-side the appropriate number of children, our list would begin:

$$0, 0, 0 \ldots \qquad 1, 1, 1 \ldots \qquad 2, 2, 2 \ldots$$
$$\underbrace{\qquad}_{\text{6 zeroes}} \qquad \underbrace{\qquad}_{\text{11 ones}} \qquad \underbrace{\qquad}_{\text{20 twos}}$$

Rather than writing down the complete list, the total number of children can be more easily found as:

$$(6 \times 0) + (11 \times 1) + (20 \times 2) + \ldots + (2 \times 6) = 160$$

The (arithmetic) mean number of children per family is therefore

$$160/66 = 2\cdot424.$$

Since there are 66 families the median is the average of the 33rd and 34th when arranged in ascending order by number of children. Both these rankings have 2 children, giving a median of 2.

The mode is the most frequently occurring number of children per family, *i.e.* 20.

2. The annual rates of inflation in Veneguay over the last 5 years have been 8%, 11%, 13%, 18%, and 23% respectively. Find the average annual rate of inflation over this period.

Answer

The geometric average is the only really sensible measure to use in this growth situation, since the arithmetic average of

$$\frac{8 + 11 + 13 + 18 + 23\%}{5} = 14\cdot6\%$$

is not very helpful when effects are compounding year by year. The distinction between the two averages is analogous to the difference between simple and compound interest.

$$\text{The geometric mean} = \sqrt[5]{(8 \times 11 \times 13 \times 18 \times 23)}\%$$
$$= \sqrt[5]{(473616)}\%$$
$$= 13\cdot65\%$$

Exercises on Section 6

7. Eight comptometer operators are given the identical costing calculation to perform. The results they obtain, in pence, are:

$$63 \quad 63 \quad 59 \quad 65 \quad 66 \quad 66 \quad 63 \quad 62$$

Find the mean, median, and mode of these figures. Which measure of centrality would you use to indicate most closely the true value of the cost?

8. The number of flaws occurring per 10 metres of cloth has, over the past 100 metres, been:

$$1 \quad 4 \quad 2 \quad 6 \quad 2 \quad 1 \quad 3 \quad 5 \quad 2 \quad 7$$

Find the mean, mode, and median and explain their significance verbally.

9. The hourly rates of pay received by electrical workers at Transistorite Limited are as follows:

(*a*) Standard rate 90.12p/hour.
(*b*) Normal overtime rate 135.20p/hour.
(*c*) Sunday rate 180.24p/hour.
(*d*) Special Christmas/New Year/Easter rate 225.24p/hour.

The total numbers of hours paid for by the company in 1974 in each of the four categories are (*a*) 1320 hours, (*b*) 406 hours, (*c*) 210 hours, (*d*) 124 hours.

Calculate the average hourly rate paid by Transistorite Limited.

10. Table 8 gives the annual profits before tax of a company over a period of 10 years. Find the average profits over each consecutive 5-year period.

Table 8. Annual profits 1963–1972

Year	Profits (£000)
1963	100
1964	103
1965	115
1966	110
1967	119
1968	60
1969	72
1970	100
1971	180
1972	175

7. Mean of grouped data. It may be the case that the original detailed information is not readily available to calculate a measure of centrality. Alternatively, the sheer bulk of the data may make it laborious or impractical to examine every unit of data. Consequently the raw data is often presented in the form of a frequency table. How may we calculate an approximate measure of centrality in these circumstances?

In the following, attention is directed towards the arithmetic mean only. Although methods are available for estimating the median and mode for grouped data it is doubtful whether the results are particularly useful.

Suppose that the daily takings of the preceding section were given in the manner of Table 9.

Table 9. Distribution of daily takings

Daily takings, T, in £	Number of days
$0 \leqslant T < 100$	10
$100 \leqslant T < 200$	30
$200 \leqslant T < 300$	50
$300 \leqslant T < 400$	80
$400 \leqslant T < 500$	100
$500 \leqslant T < 600$	85
$600 \leqslant T < 700$	75
$700 \leqslant T < 800$	40
$800 \leqslant T < 900$	25
$900 \leqslant T < 1000$	5

Although the exact value of the mean of the 500 takings cannot be found without recourse to the original data, an approximate value known as the *grouped mean* or the *mean of the grouped data* can be found readily. We assume that in each class interval the distribution of values above and below the midpoint of the interval is roughly equal. Thus, in the interval £0–£100 an approximate value for the total of the 10 days' takings is $10 \times £50 = £500$. Proceeding in a similar way through the remaining 9 class intervals, the total of the 500 takings is approximated by:

$$(10 \times £50) + (30 \times £150) + (50 \times £250) + \ldots$$
$$+ (5 \times £950) = £242{,}000$$

To find the grouped mean all that is necessary is now to divide this total by 500:

$$\text{Grouped mean} = \frac{£242{,}000}{500}$$
$$= £484$$

More generally, if the r class intervals are numbered $i = 1, \ldots, r$ whilst m_i, f_i denote the midpoint of, and number of measurements in, the ith interval respectively, then the grouped mean is:

$$\left(\sum_{i=1}^{n} f_i m_i \right) \Big/ \left(\sum_{i=1}^{n} f_i \right)$$

Alternatively, if n is the total number of measurements, the grouped mean is:

$$\left(\sum_{i=1}^{n} f_i m_i \right) \Big/ n$$

Exercises on Section 7

11. Find the grouped mean of the data given in Table 10 which give the price of a certain quoted share over a period of 200 days.

Table 10. Price of a share over 200 days

Share price (S) in pence	Number of days
$100 < S \leqslant 105$	5
$105 < S \leqslant 110$	8
$110 < S \leqslant 112$	14
$112 < S \leqslant 114$	30
$114 < S \leqslant 116$	47
$116 < S \leqslant 118$	38
$118 < S \leqslant 120$	36
$120 < S \leqslant 125$	18
$125 < S \leqslant 130$	4

12. It is often simpler to code the data when calculating summary statistics. In Question 11, for example, the mean might be guessed to have a value near 115. Accordingly, the class intervals can be expressed in terms of deviations from this value giving a revised presentation (Table 11). Calculate

Table 11. Coded share prices from Table 10

(Share price −115) = S'	Number of days
$-15 < S' \leqslant -10$	5
$-10 < S' \leqslant -5$	8
$-5 < S' \leqslant -3$	14
$-3 < S' \leqslant -1$	30
$-1 < S' \leqslant 1$	47
$1 < S' \leqslant 3$	38
$3 < S' \leqslant 5$	36
$5 < S' \leqslant 10$	18
$10 < S' \leqslant 15$	4

the grouped mean for this table, and use it to find the grouped mean of the uncoded data.

8. Dispersion. Unfortunately a measure of centrality alone does not provide a sufficiently adequate summary of a set of figures. Consider the two sets

(a) −2, −1, 0, 0, 1, 2
(b)   0,   0, 0, 0, 0, 0

In both cases the mean, median, and mode are all zero. The difference in character between the two sets lies not in their centrality, but in their variation about a central value.

As in the measurement of central tendency, so in the measurement of dispersion several statistics are available. Of these the standard deviation and variance are by far the most commonly used, and will be considered in some detail in a moment. Briefly some of the others, together with their method of calculation, are described below in the context of the data given in Table 12. We are interested in particular in the dispersion of the figures in the last column of the table.

The difference between the highest and lowest measurements, is the *range*, e.g. $3 \cdot 2 - (-1 \cdot 3) = 4 \cdot 5$.

To obtain the *interquartile range* firstly find the median of the distribution. Secondly find the medians of the upper and lower halves of the distribution. These are called respectively the *upper* and *lower quartiles* since one-quarter of the distribution lies above the former and one-quarter below the latter. The range between the upper and lower quartiles is the interquartile range.

Table 12. Budget variances

Budget	Budgeted figure (£000)	Actual figure (£000)	Differences (£000)
A	10·3	8·9	1·4
B	11·2	12·5	−1·3
C	21·3	18·1	3·2
D	4·8	6·0	−1·2
E	18·9	19·1	−0·2
F	9·1	10·1	−1·0
G	7·4	7·3	0·1
H	3·2	2·8	0·4
I	15·1	16·4	−1·3
J	10·2	9·0	1·2
K	5·4	6·3	−0·9
L	6·8	6·0	0·8
M	13·9	14·8	−0·9
N	7·0	5·9	1·1

e.g. arranging the figures in ascending order of magnitude we have:

$$-1\cdot3, \ -1\cdot3, \ -1\cdot2, \ -1\cdot0, \ -0\cdot9, \ -0\cdot9, \ -0\cdot2, 0\cdot1, 0\cdot4,$$
$$0\cdot8, 1\cdot1, 1\cdot2, 1\cdot4, 3\cdot2.$$

The median is

$$\frac{-0\cdot2 + 0\cdot1}{2} = -0\cdot05,$$

and the upper and lower quartiles are 1·1 and −1·0, respectively. The interquartile range is, therefore, $1\cdot1 - (-1\cdot0) = 2\cdot1$.

The average value of the deviations (irrespective of sign) of all the measurements from the mean is the *average (or mean) absolute deviation.*

e.g. the mean value of the figures in the last column is

$$\frac{(1\cdot4) + (-1\cdot3) + \ldots + (-0\cdot9) + (1\cdot1)}{14} = 0\cdot1$$

Working down the column the absolute deviations from the mean are respectively:

$$1\cdot3, \ 1\cdot4, \ 3\cdot1, \ 1\cdot3, \ 0\cdot3, \ 1\cdot1, \ 0\cdot0, \ 0\cdot3, \ 1\cdot4, \ 1\cdot1, \ 1\cdot0, \ 0\cdot7, \ 1\cdot0, \ 1\cdot0.$$

Finally, the average absolute deviation is

$$\frac{1\cdot3 + \ldots + 1\cdot0}{14} = \frac{15\cdot0}{14} = 1\cdot07$$

The range and interquartile range have the advantage that they are easy to calculate. Both can be, however, rather misleading measures of dispersion. Thus the ranges of the two sets

(a) 0, 0, 0, 0, 0, 0, 0, 0, 10
(b) 0, 0, 0, 0, 0, 10, 10, 10, 10, 10

are both 10. Similarly, the two sets

(a) −800, −600, −400, 1, 2, 2, 3, 4, 5, 7, 9, 200, 400, 600
(b) 0, 1, 1, 1, 2, 2, 3, 4, 5, 7, 9, 9, 10, 10

have the same interquartile range of 8.

In general the interquartile range is more useful than the range since it is not sensitive to outlying extreme values. The average absolute deviation has none of the defects of the other two measures since it takes account of the distance of each and every figure from a central value. Unfortunately it is a difficult expression to deal with algebraically in any development of statistical theory. In later work it will be necessary to combine distributions, for instance, and there is no simple way of expressing the average absolute deviation of a composite distribution in terms of those of its component distributions.

9. Standard deviation and variance. Turning now to the standard deviation, its method of calculation is illustrated using the last column of Table 12.

Data:

1·4, −1·3, 3·2, −1·2, −0·2, −1·0, 0·1, 0·4, −1·3, 1·2, −0·9,
0·8, −0·9, 1·1.

(a) Find mean of figures (this provides the "centre" about which dispersion is measured):

$$\frac{1·4 + \ldots + 1·1}{14} = 0·1$$

(b) Find the deviation of each figure from the mean:

1·3, 1·4, 3·1, −1·3, −0·3, −1·1, 0·0, 0·3, −1·4, 1·1, −1·0,
0·7, −1·0, 1·0.

(c) Square the deviations found in (b). (The result of simply adding together the deviations in (b) would be zero, because of the way in which the mean is defined. The effect of squaring the deviations is to make them all positive to prevent this cancellation):

1·69, 1·96, 9·61, 1·69, 0·09, 1·21, 0·00, 0·09, 1·96, 1·21, 1·00,

0·49, 1·00, 1·00.

(d) Sum the squared deviations in (c):

$$1 \cdot 69 + \ldots + 1 \cdot 00 = 23 \cdot 00$$

(e) Divide the total in (d) by the number of measurements. (The resulting figure is known as the *variance*, and is the average value of the squared deviations. Since this figure is representative of the deviation of a single measurement from the "centre," and is therefore not reflecting the number of original measurements in any way, the standard deviation has the important property that it can be used for comparing the dispersion of different-sized sets of measurements):

$$\frac{23 \cdot 00}{14} = 1 \cdot 64$$

(f) Take the square root of the result in (e). (The reason for this step is to reduce the "distorting" effect introduced at (c), and to provide a measure of dispersion which is in the same units as the original data):

$$\sqrt{(1 \cdot 64)} = 1 \cdot 28 = \text{standard deviation (S.D.)}$$

In general terms, if the original data contain n figures x_i ($i = 1, \ldots, n$), the standard deviation (s) is given by the formula:

$$s = \sqrt{\left\{ \sum_{i=1}^{n} \frac{(x_i - \bar{x})^2}{n} \right\}}$$

where

$$\bar{x} = \text{mean} = \left(\sum_{i=1}^{n} x_i \right) \Big/ n$$

Computationally, the standard deviation is obtained more easily from the formula:

$$s = \sqrt{\left\{ \frac{\sum_{i=1}^{n} x_i^2}{n} - \left(\frac{\sum_{i=1}^{n} x_i}{n} \right)^2 \right\}}$$

Perhaps the best way of remembering this expression is verbally—the variance (*i.e.*, s^2) is the average value of the squares, less the square of the average value.

Exercises on Section 9

13. Calculate the range, interquartile range, average absolute deviation, and standard deviation of the "budgeted figures" in Table 12.
14. Find the variance in the "actual figures" in Table 12.
15. Check that the two formulae for calculating standard deviation give the same result, using the profit figures in Table 8.

10. Standard deviation for grouped data. For the reasons given in Section 7 above it is convenient to have a method for calculating an approximate value for the standard deviation when a frequency table is provided. Using the example in Section 7, Table 9 is extended by two further columns (*see* Table 13).

Table 13. Daily takings

Takings, T, in £	Number of days (f_i)	Midpoint of interval (m_i)	Deviation of midpoint from group mean	Squared deviation of midpoint from mean
$0 \leqslant T < 100$	10	50	−434	188,356
$100 \leqslant T < 200$	30	150	−334	111,556
$200 \leqslant T < 300$	50	250	−234	54,756
$300 \leqslant T < 400$	80	350	−134	17,956
$400 \leqslant T < 500$	100	450	−34	1156
$500 \leqslant T < 600$	85	550	66	4356
$600 \leqslant T < 700$	75	650	166	27,556
$700 \leqslant T < 800$	40	750	266	70,756
$800 \leqslant T < 900$	25	850	366	133,956
$900 \leqslant T < 1000$	5	950	466	217,156

The fourth column gives the distance of the midpoint of the interval from the group mean, £484, and the fifth column gives the squares of these values. In fact, these columns are derived by analogy with the stages (*b*) and (*c*) of the standard deviation calculation in Section 9 above. The argument really is that, since the exact mean is unknown, one can use the grouped mean in its place. In the absence of exact values for the 10 figures in the first class interval, they are all assumed to have a value of £50 (the mid-value). Consequently in stage (*d*) of the standard deviation calculation we have:

$$\underbrace{(-434)^2 + \ldots (-434)^2}_{10 \text{ times}} + \ldots + \underbrace{(466)^2 + \ldots + (466)^2}_{5 \text{ times}}$$

or, more concisely:

$$10(434)^2 + 30(334)^2 + \ldots + 25(366)^2 + 5(466)^2 = 19,222,000$$

Moving now to stage (*e*), leads to:

$$\frac{19,193,200}{10 + \ldots + 5} = \frac{19,222,000}{500} = 38,444$$

Finally, the grouped standard deviation obtained at stage (*f*) is:

$$\sqrt{(38,444)} = 196 \cdot 07$$

The method is conveniently summarised in the general formula:

$$\text{Grouped S.D. } (s) = \sqrt{\left\{\frac{\sum\limits_{i=1}^{n} f_i(m_i - M)^2}{\sum\limits_{i=1}^{n} f_i}\right\}}$$

where

$$M = \text{grouped mean} = \left(\sum_{i=1}^{n} f_i m_i\right) \Big/ \left(\sum_{i=1}^{n} f_i\right)$$

Once again from a computational point of view, the expression:

$$s = \sqrt{\left\{\frac{\sum\limits_{i=1}^{n} f_i m_i^2}{n} - \left(\frac{\sum\limits_{i=1}^{n} f_i m_i}{n}\right)^2\right\}}$$

is more convenient.

Exercises on Section 10

16. Calculate the grouped standard deviation for the data given in Table 13 using the alternative formula.

17. Calculate the grouped standard deviation for the data given in Table 1.

CHAPTER 2

PROBABILITY

INTRODUCTION

In every aspect of day-to-day life we are all faced with situations and decisions involving uncertainty. Will the weather be fine today? How long should I allow to travel to work? Will it be quicker by train or by car?

More specifically the company director, the investment analyst, the stockbroker, the actuary, the accountant—in fact anyone connected with decision-making in the financial and business world—all have to deal with the problem of uncertainty in their work. In view of the complexity of modern business it is not surprising that the tools of probability theory—enabling the rational assessment of the future implications of adopting a particular course of action—are becoming indispensable to the decision and policy makers.

The purpose of this chapter is to develop the fundamental notions underlying probability. Most of us, of course, have a fairly clear idea that when a coin is tossed, the probability of its landing head uppermost is $\frac{1}{2}$, or that when a die is thrown the probability of a three, say, is $\frac{1}{6}$. Each of these statements expresses our uncertainty about the future result of an action. In a similar way a book-maker uses terms like "20 to 1 against" and "2 to 1 on," to measure his uncertainty about the outcome of a race. It is not easy, however, to justify our use of statements like these, nor indeed to explain exactly what we mean by them. Even more difficult is to analyse the meaning of the following type of remark: "It is likely to rain tomorrow," or "I shall probably invest in gilts," or "We shall expect to make a profit on the sales of our new product."

By the end of the chapter we will have a clearer idea what is meant by a statement involving probabilities.

SETS

1. Definitions. As we all are aware the future can be analysed in terms of what may or may not happen. We shall be concerned with placing future happenings into categories labelled "possible," and "impossible," and with examining relationships between future happenings such as "the pound will be devalued," "our trade deficit will decrease," "company profits will increase," and so on. The relationships between future happenings can be expressed very simply in *set notation*, and we shall be concerned initially with gaining some knowledge about *sets*.

We use the concept of a set in everyday conversation. We may talk of a "a group of shareholders," "a bunch of grapes," "a collection of stamps," or "a variety of opinions." All these are examples of sets, the essential feature of which is a grouping of objects or abstract ideas which share some common property.

There are two ways of defining a particular set. In the first instance, a set may be defined simply by listing all its members. Alternatively, a set may be thought of in terms of a rule which clearly enables one to decide whether any "object" (in the broadest sense of the word) belongs to the set or not. Usually the first method is the more precise, but on the other hand, it can be quite a lengthy process. The second method needs to be used with care. Consider, for example, the simple "prescriptive" definition: "The set of all accountants." This is very concise, but unfortunately ambiguous. What in fact is meant by an accountant? Does the definition include articled clerks, or only members of professional bodies (and if so, which bodies) and does it include graduates in accounting?

Since the alternative—listing by name all those people in the world we consider to be accountants—is not very practical, the better definition involves a considerable tightening of the prescriptive rule for membership of the set.

There is not too much difficulty in doing this, but the necessity of being unambiguous if the shorter method of definition is used should now be clear.

2. Notation. It will be useful to introduce some notation at this point. Usually a set is denoted by a capital letter A, B, C, ..., simply as a shorthand device to avoid quoting the defining rule of the set each time it is mentioned. Having said, for example, that A is the set of all Associates and Fellows of the Institute of Chartered Accountants of England and Wales, in future the set can be simply referred to as A.

The statement defining A is written in the following form:

$$A \equiv \{\text{Associates and Fellows of the I.C.A.E.W.}\}$$

The use of the brackets { } to contain the listed *members* (or *elements*) of a set, or its defining rule, is fairly standard. The symbol $\equiv$ requires comment. Literally it means "identically equal to," and indicates that A and the expression in brackets are exactly the same.

Taking another example, we define set B by:

$$B \equiv \{\text{Even numbers}\}$$

(*i.e.*, B is the set of all even numbers).

One may also express B in the form of a list:

$$B \equiv \{ \ldots, -6, -4, -2, 0, 2, 4, 6, \ldots \}$$

Here, the ... at the beginning and end indicate that the listing continues in both directions. Incidentally, the two sets A and B are different in nature.

In the case of A, given sufficient patience, it would be possible to list all the members—in other words A has a *finite* number of members. On the other hand, however much patience and energy we had, we could never list all the members of B, since there is an endless sequence of numbers divisible by two. B is said to be an *infinite* set (*i.e.*, it has an infinite number of members).

Worked Examples

1. List the members of the set of countries belonging to the Common Market.

Answer
West Germany, France, Belgium, Holland, Italy, Luxembourg, Great Britain, Eire, Denmark.

2. How many members has the set $G \equiv \{$Tax payers who pay no tax$\}$?

Answer
None! This set is quite a respectable one, and is known as the *empty set*.

Exercises on Section 2

1. List the members of the set of even numbers less than 20 which are divisible by 3.

2. Which of the following sets have a finite number of members:
   (*a*) Transactions in a bank over a 5-year period.
   (*b*) The set of grains of sand on the earth.
   (*c*) The set of odd numbers divisible by 5.
   (*d*) The set of possible values of the rate of return on an investment.

3. Give an example of a set of sets.

4. Sets C, D, are defined by:
   $C \equiv \{$Debtors of a particular company$\}$,
   $D \equiv \{$Accidents involving cars insured by Rekless Insurance Co. Ltd.$\}$

Do you think these are adequate definitions, *i.e.* does each definition enable you to decide whether any particular "object" belongs to the set or not?

3. Operations on sets. Having clarified what is meant by a set, we now extend our ideas to the relationships between sets.

A great deal of simplification is brought about when dealing with sets and their relationships if we use *Venn diagrams*. These diagrams are used to give a pictorial representation of the sets, and are based on the simple notion of denoting a set by a closed curve (*see* Fig. 11 (*a*)). The examples in Fig. 11 illustrate their use. In each case the shaded area represents the named set.

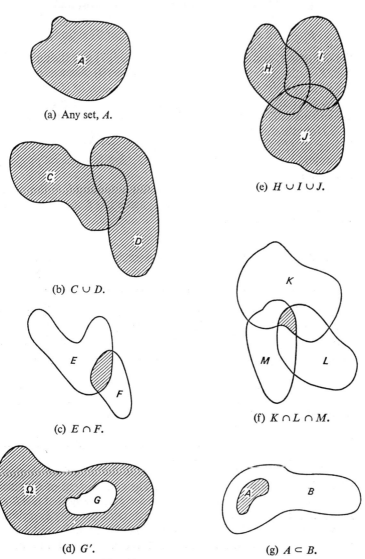

(a) Any set, A.

(b) $C \cup D$.

(c) $E \cap F$.

(d) G'.

(e) $H \cup I \cup J$.

(f) $K \cap L \cap M$.

(g) $A \subset B$.

FIG. 11.—Relations between sets.

Following normal convention, the symbol Ω, (Greek: omega) in the fourth example denotes the "universe."

(*a*) The *union* of two sets is the set that contains all the elements which are in either (or both) of the two sets. For example, if

$$C \equiv \{1, 2, 3, 4, 5\}, \quad D \equiv \{1, 2, 6, 7\},$$
$E \equiv \{\text{All Associates of the Institute of Cost and}$
$\qquad \text{Management Accountants}\},$
$F \equiv \{\text{All Associates of the Institute of Taxation}\},$

the union of C and D is the set $\{1, 2, 3, 4, 5, 6, 7\}$ and the union of E and F is the set of persons who are *either* (*i*) Associates of the Institute of Cost and Management Accountants, *or* (*ii*) Associates of the Institute of Taxation, or both. In symbols the combined sets are written as $C \cup D$ and $E \cup F$ (read "C union D," and "E union F"). Note that members belonging to *both* C and D automatically belong to $C \cup D$, and that persons who are Associates of *both* Institutes automatically belong to $E \cup F$. This is a general property of a union.

(*b*) The *intersection* of two sets on the other hand, is the set of elements that belong to both sets. This again has a special notation, $\cap$, and in the examples,

$C \cap D \equiv \{1, 2\}$
$E \cap F \equiv \{\text{All persons who are Associates both of the}$
$\qquad\qquad\text{I.C.M.A. and of the Institute of Taxation}\}$
$\qquad$ (read "C intersection D," and "E intersection F").

(*c*) A *subset* is a set which is completely contained within another set. If $G \equiv \{1, 2, 3\}$, then G is a subset of C, since every element of G is contained in C. This special relationship between G and C is expressed as $G \subset C$ (read "G is contained in C," or "G is a subset of C").

Note that G is not a subset of D, because the element 3 belongs to G, but not to D. For a set to be a subset of another set, the former must be wholly contained in the latter.

Incidentally, as a consequence of definitions (*a*), (*b*), and (*c*) the intersection of any two sets A and B is a subset of the union of the two sets, *i.e.* $(A \cap B) \subset (A \cup B)$.

(*d*) The *complement* of any given set is the set of all elements which do not belong to the given set. The complement of E, which is written E', is the set whose elements are everything that is *not* an A.C.M.A. It is a peculiar set in that it contains not only everybody who is not an A.C.M.A., but everything else as well—red buses, books on statistics, the moon, Marxism, and so on! The list is endless.

Expressions involving more than two sets may be interpreted in a natural extension of the ideas that have been introduced. Thus $A \cup B \cup C$ is the union of A, B, and C, *i.e.* the set of all members of either A, or B, or C.

Note however that the expression $A \cup B \cap C$ is ambiguous, since it may mean either (*a*) the intersection of $A \cup B$ with C, or (*b*) the union of A with $B \cap C$. To remove the ambiguity in an expression involving both union and intersection signs, it is necessary to bracket some of the terms together. Thus we may rewrite (*a*) and (*b*) more clearly as:

(*a*) $(A \cup B) \cap C$
(*b*) $A \cup (B \cap C)$ (*see* Fig. 12).

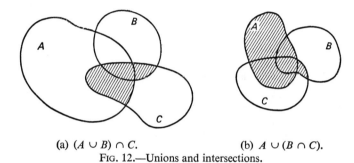

(a) $(A \cup B) \cap C$.   (b) $A \cup (B \cap C)$.
Fig. 12.—Unions and intersections.

Worked Examples

1. Using Venn diagrams illustrate that $(A \cap B)' = A' \cup B'$.
Is this equation true for any two sets A and B? Interpret the equation verbally.

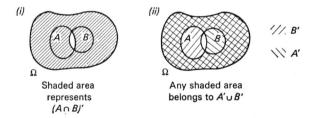

(i)   *(ii)*   ///. B'
                                                  \\\ A'

Shaded area   Any shaded area
represents   belongs to $A' \cup B'$
$(A \cap B)'$

Answer

The two sets in the Venn diagram could be any sets, and the diagram above represents the quite general relationship between two sets. The equation is therefore always true. Verbally: Something which does not belong to both A and B, either does not belong to A or does not belong to B.

2. A company manufactures three products—locks, stocks, and barrels. The products require machining in the course of their manufacture. It is planned to increase the output of the highly profitable stocks, to the extent that 60 machines will be needed to work full-time on machining stocks.

The machines presently operated by the company are such that:

15 machines can be used for machining at least stocks and barrels;
8 machines can be used for machining at least barrels and locks;
30 machines can be used for machining at least locks and stocks;
10 machines can be used for machining stocks only;
3 machines can be used for machining barrels and locks, but not stocks.

Assuming that production of locks and barrels can be reduced to free any existing machine exclusively for stock production, how many new machines will it be necessary to purchase? Use Venn diagrams.

Answer

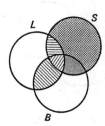

We denote machines which can be used for making stocks by *S*, etc.
There are 8 machines in *L*∩*B*, and 3 of these are in the shaded (\\\\\\) area. There are therefore 5 machines in *L*∩*S*∩*B* (*i.e.,* area ///).
Now *L*∩*S* contains 30 machines, 5 of which belong to *L*∩*S*∩*B*. Thus area ≡ contains 25 machines.
Similarly *S*∩*B* contains 15 machines, 5 of which belong to *L*∩*S*∩*B*.
Thus area ▓ contains 10 machines.

Finally, we are told that area

contains 10 machines only. The number of machines available for working on stocks is the sum of the numbers of machines in areas

///, ≡, ▓, ▨,

i.e. 5 + 25 + 10 + 10 = 50. It will be necessary to purchase a further 10 machines for working on stocks.

Exercises on Section 3

Define *A, B, C* as follows:

$A \equiv$ {Members of the I.C.M.A.},
$B \equiv$ {Graduates of British Universities},
$C \equiv$ {Members of the A.C.C.A.}.

The three sets may be represented in the form of a Venn diagram (Fig. 13).

5. Suggest a suitable universe set, Ω.

6. Locate the following sets in Fig. 13:

(a) $A \cap B$ (e) $A \cap C''$
(b) $(A \cap B)'$ (f) $A \cup C$
(c) $(C \cup B)'$ (g) $C \cup B'$
(d) $(C' \cap B)$ (h) $C' \cup B$

Describe verbally the sets (a)–(h).

7. Interpret verbally the regions numbered 1–8 in Fig. 13.

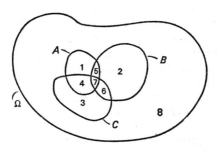

FIG. 13.—Venn diagram for Exercises on Section 3.

8. The following sets represent the regions 1–8, but not in order. Identify which set corresponds to which region

$A \cap B \cap C'$, $A \cap B \cap C$, $(A \cup B)' \cap C$, $(B \cup C)' \cap A$,
$B \cap C \cap A'$, $(A \cup C)' \cap B$, $(A \cup B \cup C)'$, $A \cap C \cap B'$

9. Give the set notation in terms of A, B, C for

(a) The combined region made by regions 6 and 7.
(b) The combined region made by regions 5 and 7.
(c) The combined region made by regions 1, 3, 4, 5, 6, 7.

10. Identify the following regions:
(a) Graduates of British Universities who are neither members of the I.C.M.A., nor members of the A.C.C.A.
(b) Graduates who are members of the I.C.M.A. but not of the A.C.C.A.
(c) Individuals who are not graduates, not members of the I.C.M.A., and not members of the A.C.C.A.

11. Using Venn diagrams illustrate that:

$A \cup (B \cap C) = (A \cup B) \cap (A \cup C)$
Interpret the relationship verbally.

Do you think the relationship is true for any sets A, B, and C?

EXPERIMENTS AND EVENTS

4. Introduction. We are now nearly in a position to link our knowledge of sets with the idea of probability.

In its simplest terms probability provides a quantitative measure of the degree of uncertainty about occurrences and situations. As such, a probability statement may relate to an incident which has not yet happened, or to an incident which has already occurred, the result of which is not yet known. In both cases the uncertainty, and therefore the statement of probability, can be seen to arise from a lack of complete information about the occurrence and its outcome. Implicit in these comments is the assumption that there are several possible outcomes to the incident. Clearly, if there were only one possible outcome, then the introduction of probability would be unnecessary. If, for example, a man were to claim allowances for 60 children on his annual tax return, it is certain that the outcome would be an investigation by the Inspector of Taxes!

5. The random experiment. The scientific meaning of an experiment is essentially that of conditions that are controlled (and that can be replicated at will), together with observation of some happening under those conditions. A chemist knows, for example, what the result will always be of mixing two chemicals under the same conditions. In statistical work, however, the term is used in a more general sense and includes, for example, tosses of a coin, drawing of cards from a pack, drawing a random sample from a large number of accounts receivable, and selecting a random portfolio of shares. Each of these is a controlled set of conditions, but notice that in any particular trial of the experiment there is an element of uncertainty in that one of several outcomes may result.

In the case of the coin, the experiment is controlled to the extent that on each trial a coin is made available, together with a mechanism for tossing it. However, there the control ends. (If the coin could be tossed under *perfectly* replicable conditions it would, of course, land with the same face uppermost each time.) Similarly, the random selection of a sample of accounts receivable, or of a portfolio of shares, are both subject to an element of uncertainty, in that repetitions of the sampling will result in differing samples being drawn. Accordingly, a statistical experiment is called a *random experiment*. Looking at this another way, a random experiment is a situation which can lead to any one of several possible outcomes.

6. Sample spaces. A special name—the *sample space*—is given to the set of possible outcomes of a random experiment. In the case of tossing a coin, for instance, there are two possible results of a single trial—

head (H), or tail (T) (the possibility of the coin landing on its edge is excluded from the following discussion), and therefore two elements in the sample space. If now a random experiment is defined as three successive tosses of the coin, the sample space will have eight elements: *HHT, HTT, HTH, HHH, THT, TTT, TTH, THH*, which represent all the possible outcomes of the three successive tosses. Of these elements one, and only one, will actually occur.

As a further illustration, consider a simple portfolio of three shares 1, 2, 3 in different quoted companies. On any day the market price of each share can either increase (I), decrease (D), or remain unchanged (U). (For simplicity we exclude the possibilities of takeovers, or suspension in dealings.) The behaviour of the portfolio during the day can be seen (at the beginning of the day) as a random experiment. The sample space can be displayed as a Venn diagram (Fig. 14).

DDD	*III*	*UUU*
DDI	*IID*	*UUD*
DDU	*IIU*	*UUI*
DUD	*IUI*	*UDU*
DUI	*IUD*	*UDD*
DUU	*IUU*	*UDI*
DID	*IDI*	*UIU*
DII	*IDD*	*UID*
DIU	*IDU*	*UII*

Fig. 14.—Venn diagram representing the set of 27 possible behaviours of three shares in a day's dealings.

Each element (or member) of a sample space is referred to as an *elementary event*. Equivalently, each possible outcome of a random experiment is an elementary event. In the last example, the sample space contains 27 elementary events.

7. Events. Since a sample space is simply a special type of set, whose elements are the possible outcomes of a random experiment, it is quite sensible to talk about subsets of the sample space. Thus {*DDD, III, UUU*}, {*IUD, IDI, UII, UUU, DUI*} and {*III, IID, IIU*} are all subsets of the sample space in the previous example. In the coin example {*THT, HHH*}, and {*THT, THH, HHT, HHH*} are two subsets in the sample space.

Any subset of a sample space—such as those listed above—is called an *event*.

The use of the term *elementary event* to refer to a single possible outcome is reasonable in that it is the "smallest" type of subset in the sample space. One may wonder, however, at the definition of an event as a set of possible outcomes, since only one outcome will actually occur. The point here is that, whilst it is true, for example, that on a particular day only one pattern of behaviour of the share prices can take place, the view of the random experiment is taken from beforehand. In other words, the portfolio's behaviour is being considered at the beginning of the day looking forwards, and not at the end of the day looking back. Event L, defined by $L \equiv \{III, IID, IIU\}$ is the event "shares A, B increase in value." It contains three members, since there are three ways in which L may be fulfilled. $M \equiv \{DDD, III, UUU\}$ is the event "all three shares behave the same."

These events, and indeed all other subsets of the sample space are perfectly meaningful. Similarly, in the coin example, it is quite meaningful to talk of the event "heads on second toss."

This event is built from the four elementary events THT, THH, HHT, HHH, of which at most one (and possibly none) may occur.

The previous few paragraphs contain a number of unusual terms, and it may be helpful to summarise them in the following manner:

General terminology	Equivalent special terms associated with random experiments
Set	Sample space
Element/member	Elementary event/outcome of random experiment
Subset	Event

It is important that the reader has a firm grasp of the ideas and terms that have been introduced in this section. The concepts are not difficult, but they are almost certainly unfamiliar. The following exercises should aid in the familiarisation process.

Worked Example

Three customers' accounts are to be selected from a batch of six accounts for purposes of checking.

    (*a*) Describe the sample space clearly.

    (*b*) Would the sample space be the same if we were concerned not only with which three accounts are selected, but with the order in which they are drawn?

Answer

For convenience let us label the six accounts *A, B, C, D, E,* and *F.* The sample space will consist of all the possible selections of three accounts which may be drawn out of the six, thus:

ABC	*ABD*	*ABE*	*ABF*
ACD	*ACE*	*ACF*	*ADE*
ADF	*AEF*	*BCD*	*BCE*
BCF	*BDE*	*BDF*	*BEF*
CDE	*CDF*	*DEF*	*CEF*

Since there are 20 ways in which the three accounts can be selected out of the six, the sample space of possible outcomes of the random experiment contains 20 members.

If we were also concerned with the order in which the accounts were drawn the sample space would be much larger. Concentrating on the previous outcome *ABC,* we see that this may have resulted from any of the following orders of drawing:

ABC, ACB, BAC, BCA, CAB, CBA

Accordingly, corresponding to *each* member of the sample space in (*a*) there will be six members in (*b*), resulting in a sample space of 120 members.

Exercises on Section 7

12. A certain type of machine has a maximum life of 10 years, and can be taken out of service and replaced at the end of any year after its installation. Assuming a new machine has just been installed, list:
    (*a*) The sample space of its service-life.
    (*b*) The elementary events contained in the event: "its service-life is not more than four years."
    (*c*) The elementary events contained in the event "its service-life is not less than four years."
    What is the random experiment in this example?

13. In the worked example what would be the sample space in (*a*) and (*b*) respectively, if we were told that the first account had already been drawn, and was in fact *E*?

14. An investor is to select three out of four shares 1, 2, 3, 4.
    (*a*) Define the sample space.
    (*b*) Show the following events in your sample space:

    (*i*) 4 is chosen.
    (*ii*) 3, 4 are chosen.
    (*iii*) 2, 3, 4 are chosen.

15. A research and development project will result in net profits during the next five years of not less than £20,000 and not more than £40,000. Profits

are measured in units of £1000. List the sample space of outcomes of the investment, and indicate the following events:

   (a) Profits will be not more than £30,000 and not less than £22,000.
   (b) Profits will be less than £32,000 or greater than £35,000.
   (c) Profits will be within the ranges £21,000–£25,000 or £23,000–£34,000.

EVALUATION OF PROBABILITIES

8. Relative frequency. Having now laid our groundwork it is time to consider the assignation of numerical values to the probability of an event happening.

Suppose we throw a 6-sided die. Each throw is a random experiment with 6 possible elementary events in the sample space. The experiment can be repeated many times and a record kept of the outcome of each throw. Suppose out of 1200 throws, 4 comes uppermost 210 times. *The relative frequency of occurrence* of the 4 is found by calculating the ratio $\frac{210}{1200} = 0.175$, and this value provides an estimate for the probability of any particular throw resulting in a 4.

Intuitively, one would expect that the larger the number of throws made, the closer the relative frequency of a 4 would be to the value of 1/6. In other words, the relative frequency of occurrence of a 4 tends to a "true" value of the probability of obtaining a 4 on any particular throw of the die. Graphically, the result of a long series of throws might take the form of Fig. 15.

The figure is drawn on the assumption that the die is unbiased, in

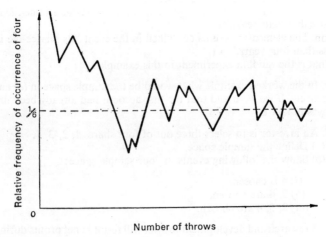

FIG. 15.—Typical pattern of result of throwing die.

the sense that no face will come uppermost more often than any other in the long run. In the case of an unbiased die the probability of any particular number resulting from a throw is 1/6.

The relative frequency approach illustrates five important desirable properties of probabilities:

(a) A probability cannot be negative, since the ratio of the number of occurrences of an event to the total number of trials cannot be less than 0.

(b) From the manner in which probability has been defined—as the limit of a relative frequency—its value, p, must be such that

$$0 \leqslant p \leqslant 1.$$

(c) The sum of the probabilities of all the possible elementary events is 1 (e.g., $1/6 + 1/6 + 1/6 + 1/6 + 1/6 + 1/6 = 1$) since, by definition, the sum of the corresponding relative frequencies is 1.

(d) The probability of an impossible event is 0 (since it never occurs, its relative frequency is 0).

(e) The probability of a certain event is 1 (since it occurs at every trial, its relative frequency of occurrence is 1).

These comments do not prove anything about probability, but they do indicate that, for probability ideas to be consistent with relative frequency ideas, these restrictions on the values should apply.

In much practical statistical work relative frequency and probability are used almost interchangeably. There are dangers in doing this, however, and it is as well to bear in mind that probability is an "ideal" value, and that relative frequency only provides a means of estimating that value.

9. Degree of belief. A serious objection to the relative frequency estimation of probabilities is that it assumes a random experiment is always repeatable under the same conditions. Unfortunately, this is not always the case. There are many situations in which measurement of the degree of uncertainty of an outcome is desirable, but where the "experiment" is of a once-and-for-all nature. A manager may wish to know what the probability will be that investment in a particular development project results in an eventual profit. Under a pre-chosen method of appraisal the possible outcomes of investment will be profit, break-even, or loss. The manager cannot, however, repeat the "experiment" of investing in the project—for obvious reasons—and calculate a relative frequency of profit. Similarly, an insurance company cannot have an individual die repeatedly in order to assess a value for the premium on a life policy!

However, it would be a sad state of affairs if the manager did not have a great deal of information, experience, and knowledge of similar situations in the past, of the state of the economy, of competitors' activities, and of the marketability of the product resulting from investment in the project.

Likewise the insurance company will have a medical report on the individual, data on mortality in the relevant age group, and information on whether the prospective policy-holder engages in dangerous pursuits.

In both cases, and indeed more generally, it is possible to pool the information available, and to produce a subjective measure of probability. Alternatively, the measure of uncertainty may be considered as a degree of belief (on a scale from 0 to 1) in an event taking place.

A great deal has been written concerning the measurement of degrees of belief. To discuss the matter thoroughly requires concepts that cannot be introduced appropriately at this stage, and so the subject is deferred to Chapter 13.

Although degree of belief and relative frequency are rather different concepts, both are attempts to assign a probability value to an event. The relationship between them can be likened to that between an aircraft and a motor car, which each use a different method for achieving the common objective of transportation.

10. The probability of an elementary event. Returning to our example concerning the portfolio of three shares in Section 6 above we notice that the sample space contains 27 elementary events (*see* Fig. 14). Using these as basic building blocks any events can be specified precisely and illustrated in a Venn diagram. Furthermore, the event "one of the 27 elementary events occurs" is certain, and therefore has a probability of 1. If we have no further information about the behaviour of share prices, we should have no reason to suppose that any elementary event is more likely to occur than any other.

Accordingly, we assign a probability of 1/27 to each of them. This reflects our belief that if the behaviour patterns of the 3 shares were recorded over a long period of time, the relative frequency of any particular combination would be approximately 1/27.

A crucial step in the above argument is that the elementary events are all equally likely. How important this step is can be illustrated quite simply. Suppose a man purchases a lottery ticket. The sample space of this "experiment" can be conveniently divided into two elementary events, *viz.* he wins, and he does not win. One could (and many do!) argue falsely that the probability of his winning is, therefore, $\frac{1}{2}$. The fallacy lies in the implicit assumption that the two elementary events are equally likely. A sound argument would be based on breaking down the sample space in greater detail, so that each elementary event represented a particular ticket being the winning one. Presumably each of these elementary events is equally likely, and—if there were say 1,000,000 tickets—the probability of the man winning would be 1/1,000,000.

11. The probability of an event. From considering the probabilities of elementary events, we now turn to the problem of how to use these in order to attach probabilities to larger events. For convenience, the sample space of the shares example is reproduced in Fig. 16.

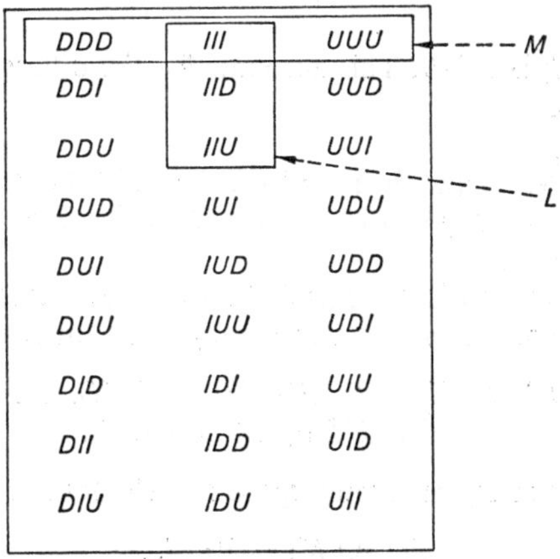

FIG. 16.—Two events concerning three-share portfolio.

Event $M \equiv$ {Shares 1, 2, 3 behave similarly} occurs if any one of the three elementary events III, DDD, UUU occurs.

Event $L \equiv$ {Shares 1 and 2 increase in value} occurs if any one of the three elementary events IID, III, IIU occurs. Since each of these occurs with a probability of 1/27, the probability that L occurs, $P(L)$, is $3/27 = 1/9$.

From a relative frequency point of view one of these three elementary events occurs 1 out of 9 times in a long run. Likewise, if $N \equiv$ {none of the share prices increases}, this event is seen to contain the 8 elementary events in Fig. 17. Accordingly

$$P(N) = 8/27.$$

The evaluation of probabilities of more complicated events presents little difficulty. Consider

$$K \equiv \text{\{any two of the prices do not increase,}$$
$$\text{whilst the third remains unchanged\}.}$$

This event is made up of the elementary events DDU, DUD, DUU, UDD, UDU, UUD, UUU. Thus $P(K) = 7/27$.

In set notation $K \subset N$ and the only elementary event contained in

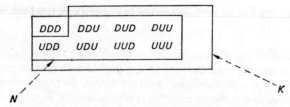

Fig. 17.—Events N and K concerning three-share portfolio.

N, but not in K, is DDD. This accounts for the difference of 1/27 in $P(N)$ and $P(K)$.

12. Compound events and their probabilities. Since events are subsets of a sample space, they can be combined in the ways that were discussed in Section 3.

It is now possible to find the probabilities of events such as $L \cup K$, $L \cap K$, and so on. Listing the elementary events in $L \cup K$ gives:

$$L \cup K \equiv \{III, IID, IIU, DDU, DUD, DUU, \\ UDD, UDU, UUD, UUU\}.$$

Since this subset of the sample space has 10 members,

$$P(L \cup K) = 10/27.$$

Verbally $L \cup K$ is a complicated event, but by using set notation, and the elementary events as basic building blocks, the analysis results in a simple calculation to find its probability.

Defining $J \equiv \{$share 3 does not increase in price$\}$, $P(J)$ is found to be $18/27 = 2/3$. The calculation of $P(J \cup L)$, however, introduces a new feature. Unlike the pair K and L, the 2 events J and L have elementary events (*viz. IID* and *IIU*) in common.

In terms of Venn diagrams their relationship is shown in Fig. 18. As a result, in $J \cup L$ there are not $18 + 3 = 21$ elementary events, but only 19, giving a value for $P(J \cup L)$ of 19/27. The figure of 19 is obtained by subtracting from 21 the number of elementary events in $J \cap L$. The latter is an event, since it is a subset of the sample space, and in fact $P(J \cap L)$ is 2/27. It is now possible to state the following relationship:

$$P(J \cup L) = P(J) + P(L) - P(J \cap L) \qquad (i)$$

In terms of elementary events (which all have equal probabilities of 1/27 in this particular case) the equation may be written as

$$19 = 18 + 3 - 2.$$

This relationship is an important one, and of quite general validity

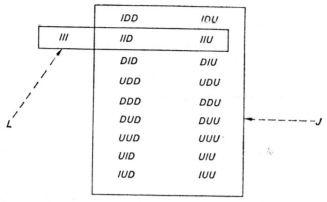

FIG. 18.—Events L and J concerning three-share portfolio.

for any pair of events in a sample space, even if the elementary events do not all have equal probabilities. As a further illustration consider the pair of events N and K.

In an exactly similar way to (i) above:

$$P(K \cup N) = P(K) + P(N) - P(K \cap N).$$

It has already been shown that $P(K) = 7/27$, and $P(N) = 8/27$. Furthermore $K \subset N$ (Fig. 17).

The event $K \cap N$ is just K itself, so $P(K \cap N) = P(K)$, and the equation for $P(K \cup N)$ becomes

$$P(K \cup N) = P(K) + P(N) - P(K) = P(N).$$

Alternatively, one could argue that since K is completely contained in N, $K \cup N$ is just N itself, and therefore

$$P(K \cup N) = P(N)$$

as before.

This result is logical, in that when K happens N will also happen.

To specify that event "K or N" will happen involves no other restrictions on the outcomes than those arising from specifying that event N will happen.

The reader may well wonder why this relationship was not used in calculating $P(L \cup K)$ earlier. In fact it was, for

$$P(L \cup K) = P(L) + P(K) - P(L \cap K)$$

and $L \cap K$ has no members, implying that the event $L \cap K$ is an impossible event. This is obvious, since the definitions of L and K clearly prevent them both happening at the same time. Thus the occurrence of L implies that both shares A and B have increased in price. This immediately precludes the occurrence of K since this involves two prices not increasing. The probability of an impossible event is 0, and therefore

$$P(L \cup K) = P(L) + P(K)$$
$$= 3/27 + 7/27$$
$$= 10/27$$

which agrees with the previous result.

Worked Example

If A, B, C are any three events show, by considering the Venn diagram of Fig. 19, that

$$P(A \cup B \cup C) = P(A) + P(B) + P(C) - P(A \cap B) -$$
$$P(N \cap C) - P(C \cap A) + P(A \cap B \cap C).$$

Hence find $P(J \cup M \cup N)$ in Section 12 above.

Answer

It is true to say that in many ways probability and area are analogous. This is perhaps most apparent when we examine a Venn diagram such as Fig. 19. The set $A \cup B \cup C$ is represented by the area contained within the dotted line. Let us try and express this area in terms of the components. Clearly Area A + Area B + Area C is too great, since the areas common to A and B, common to B and C, and common to C and A are each counted twice.

Correction for this leads to the sum:

Area A + Area B + Area C — (Area common to A and B) —
(Area common to B and C) — (Area common to C and A).

This is still not quite the correct answer, however, since the shaded area has been counted three times (in each of Areas A, B, and C), and then subtracted three times. As a result we are left with a "hole" in the middle which

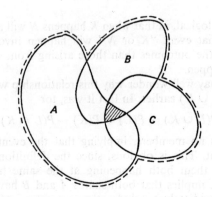

FIG. 19.—General relationship between any three sets A, B, and C.

has not been included at all. To give the final answer we need to add the area common to all of A, B, and C.

Following exactly the same line of argument in terms of probabilities rather than areas leads to the required result.

Exercises on Section 12

16. Personnel records show that, for a particular company, on average 100 people leave the service of the company each year. Of these, 80 have had more than 4 years' service, and this is the minimum time to qualify for any benefit from the pension fund.

What is the probability that a resignation will result in payment of pension benefits? Use Venn diagrams.

Of the 100 who leave per year, on average 10 return to the company's service eventually and, of these, 7 are people who had originally left after more than 4 years' service. The total work force of the company is 5000, of whom 3500 have been with the company more than 4 years.

Use relative frequency to estimate the probability in a particular year that

(*a*) a particular person leaves the company's service.

(*b*) pension benefits will be due to a particular person.

(*c*) a person who resigns will return to the company.

(*d*) a particular person who resigns will return, given that he has been with the company more than 4 years.

(*e*) a particular person will resign, given that he has been with the company more than 4 years.

(*f*) a particular person has been with the company more than 4 years, given that he has resigned.

17. The working life of a machine is not known with certainty, but the following information is available on 20 similar machines which were installed when new 5 years ago.

Life of Machine	No. of Machines
not more than 1 year	2
not more than 2 years	4
not more than 3 years	8
not more than 4 years	10
not more than 5 years	15

On the basis of this information what figure would you give for the probability that a machine will last

(*a*) more than 4 years?

(*b*) at least two years?

(*c*) between 1 and 3 years?

(*d*) more than 5 years?

18. In the example of the portfolio of 3 shares, suggest a suitable breakdown of the sample space into elementary events if prices are quoted twice in a day.

19. Find the following probabilities in Sections 11 and 12 above. For convenience the definitions of the events are repeated below:

(a) $P(L \cap M)$.
(b) $P(J \cap N)$.
(c) $P(M \cap N)$.
(d) $P(M \cap J)$.
(e) $P(L \cup M)$.
(f) $P(J \cup N)$.
(g) $P(M \cup N)$.
(h) $P(M \cup J)$.

Definitions

$J \equiv$ {Share 3 does not increase in price}.
$K \equiv$ {Any two of the prices do not increase, whilst the third remains unchanged}.
$L \equiv$ {Shares 1 and 2 increase in price}.
$M \equiv$ {Shares 1, 2, and 3 behave similarly}.
$N \equiv$ {None of prices increase}.

RELATIONS BETWEEN EVENTS

13. Complementary events. Following the discussion of compound events and their probabilities, it is appropriate to introduce at this stage some general relationships which may exist between events.

Two events are said to be *complementary* if they are "opposite" to each other. Thus "Company X will take over company Y" and "Company X will not take over company Y" are complementary events. Rather more formally, two events are complementary if one of them is bound to occur, but both cannot.

From the manner in which complementary events are defined, the sum of their probabilities must be unity. Using the special notation A' to denote the complement of event A

$$P(A) + P(A') = 1.$$

The usefulness of the concept of complementarity lies in the fact that often the calculation of the probability of an event can be shortened considerably by calculating that of its complement, and using the above relationship.

Diagrammatically, two complementary events can be represented as shown in Fig. 20.

14. Exhaustive and mutually exclusive events. The idea of complementarity leads naturally to that of exhaustive and mutually exclusive events. A group of events is *exhaustive* if they together contain all the elementary events in the sample space (*i.e.*, if their union is the whole sample space). By definition, two complementary events are exhaustive.

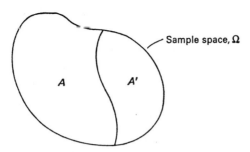

Fig. 20.—Complementary events.

If a machine has a maximum service-life of 3 years and, for simplicity, its life is measured in years, the following three events form an exhaustive set:

$A \equiv$ {Machine has a service-life of exactly 1 year}
$B \equiv$ {Machine has a service-life of exactly 2 years}
$C \equiv$ {Machine has a service-life of more than 1 year}

(We here assume that immediate obsolescence is impossible.)

This second example raises the allied notion of mutually exclusive events. A set of events is said to be *mutually exclusive* if no two of the events can occur simultaneously.

It is not difficult to see that A and B are mutually exclusive.

The 3 events A, B, and C taken together however are not, since events B and C may both occur (*i.e.*, a machine may last exactly 2 years). Clearly a pair of complementary events are mutually exclusive.

We have in fact already had an example of a set of mutually exclusive and exhaustive events—the elementary events themselves!

Consider the following situation.

Suppose a financial director of a company has been analysing the possible results of developing a new product line, and has arrived at the conclusions summarised in the following table:

Sample space of revenue (R) in £000, in first year of new product	Probabilities of these elementary events
$A \equiv \{R \geqslant 50\}$	0·20
$B \equiv \{50 > R \geqslant 40\}$	0·30
$C \equiv \{40 > R \geqslant 30\}$	0·20
$D \equiv \{30 > R \geqslant 20\}$	0·10
$E \equiv \{20 > R \geqslant 10\}$	0·10
$F \equiv \{10 > R \geqslant 0\}$	0·05
$G \equiv \{0 > R\}$	0·05

A–G represent a suitable set of elementary events, the sum of whose probabilities is therefore unity. Now the events

$$(A \cup B) \equiv \{R \geqslant 40\},$$
$$(C \cup D) \equiv \{40 > R \geqslant 20\},$$

and

$$(E \cup F \cup G) \equiv \{20 > R\}$$

are mutually exclusive and exhaustive. Furthermore,

$$P(A \cup B) = 0{\cdot}20 + 0{\cdot}30 = 0{\cdot}50,$$
$$P(C \cup D) = 0{\cdot}20 + 0{\cdot}10 = 0{\cdot}30,$$

and

$$P(E \cup F \cup G) = 0{\cdot}10 + 0{\cdot}05 + 0{\cdot}05 = 0{\cdot}20.$$

The sum of the probabilities is 1, illustrating the general fact that the sum of the probabilities of any group of mutually exclusive and exhaustive sets is always 1.

Worked Example

Are the following pairs of events complementary or not? Give reasons for your answers.

(a) I invest in Company X. I do not invest in Company Y.

(b) I shall eventually qualify as a Chartered Accountant. I shall eventually qualify as a Cost and Management Accountant.

(c) This company will make a rights issue. This company will not make a rights issue.

Answer

(a) No. I may choose to invest in X, without necessarily investing in Y.

(b) No. I may quite possibly qualify as both eventually.

(c) Yes. One of the events must occur, and both cannot.

Exercises on Section 14

20. Give several examples of sets that are

(a) Exhaustive but not mutually exclusive.
(b) Mutually exclusive but not exhaustive.
(c) Mutually exclusive and exhaustive.
(d) Neither mutually exclusive nor exhaustive.

21. Provide the complement of the following events:

(a) annual profits show a large increase;
(b) mining shares have fallen;
(c) this machine will be replaced next year;
(d) mining shares have fallen, and annual profits show a large increase;
(e) mining shares have fallen, or annual profits show a large increase.

22. An insurance company assesses the probability of a policyholder's death due to various causes as follows:

    (a) natural causes   0·95;
    (b) drowning   0·03;
    (c) motor accident   0·02;
    (d) other unnatural causes 0·04.

Do (a)–(d) form (i) a mutually exclusive group of events?

                (ii) an exhaustive group of events?

Explain why the sum of the probabilities of (a)–(d) is greater than 1.

23. In Question 22, explain why the following probabilities could not arise:

P (natural causes) = 0·95, and P (drowning) = 0·07.

(HINT: Are the two events exhaustive?)

CHAPTER 3

JOINT AND CONDITIONAL PROBABILITIES

INTRODUCTION

In the previous chapter we were concerned with establishing a basic understanding of probability. Unfortunately, in the real world events cannot always be considered in quite the simple way that we have done so far. In fact most events cannot be discussed in isolation, but only in the context of the other events which they influence, or by which they are influenced. Consequently it is often necessary to derive probabilities which reflect the dependence of events on the occurrence or non-occurrence of others.

The aim of the present chapter is to provide a framework within which such probabilities can be assigned.

JOINT PROBABILITY

1. The meaning of joint probability. Consider a firm operating a computerised accounting system in which data on customers' accounts are stored on punched cards. Suppose there are 1000 customers of whom 250 are in the industrial sector and quoted on the Stock Exchange, 300 are in the industrial sector but unquoted, 100 are non-industrial and quoted, and 350 are non-industrial and unquoted. This information is summarised in Table 14.

Table 14. Classification of 1000 shares (dependence between factors)

	Quoted (Q)	Unquoted (U)
Industrial (I)	250	300
Non-industrial (N)	100	350

Suppose a customer's account is selected at random, and interest is centred on whether the company is industrial or non-industrial. The sample space for the purpose contains just two elementary events, I (industrial) and N (non-industrial). The probabilities of these two elementary events are:

$$P(I) = (250 + 300)/1000 = 0.55$$
$$P(N) = (100 + 350)/1000 = 0.45$$

On the other hand, there may be no interest in the activities of the company, but only in whether it is quoted or unquoted. In this case, the sample space would contain the two elementary events Q (quoted) and U (unquoted) with probabilities respectively of:

$$P(Q) = (250 + 100)/1000 = 0.35$$
$$P(U) = (300 + 350)/1000 = 0.65$$

Going one step further, interest may be expressed in both the type of company, and its Stock Exchange status. Consequently, the categorisation of the result of drawing a computer card is on two levels. The outcomes classified in the two-dimensional way number four, and the probabilities attached to them are *joint probabilities*:

$$P(I, Q) = 250/1000 = 0.25$$
$$P(I, U) = 300/1000 = 0.30$$
$$P(N, Q) = 100/1000 = 0.10$$
$$P(N, U) = 350/1000 = 0.35$$

The above would be read as "the joint probability of I and Q is 0.25 etc." The idea of joint probabilities may be extended quite naturally to include situations where the classification of an outcome is at more than two levels.

In our discussion of the three-share portfolio in Chapter 2 there was essentially a three-way classification of outcomes by individual share price behaviour. Another way of expressing the probabilities is in terms of the joint probabilities of these three behaviours:

$$P(I, D, I) = 1/27$$
$$P(U, D, D) = 1/27, \text{ etc.}$$

2. Marginal probabilities and independence. Returning to the random selection of an account, suppose that the breakdown of customers had been as in Table 15.

Table 15. Classification of 1000 shares (independent factors)

	Quoted (Q)	Unquoted (U)
Industrial (*I*)	300	450
Non-industrial (*N*)	100	150

Accordingly, the joint probabilities are:

$$P(I, Q) = 300/1000 = 0\cdot30$$
$$P(N, Q) = 0\cdot10$$
$$P(I, U) = 0\cdot45$$
$$P(N, U) = 0\cdot15$$

The probabilities associated with the possible outcomes, taking account of only one-way classification, are designated *marginal probabilities*. Thus, $P(I) = 0\cdot75$ is the marginal probability of an industrial company being selected. In a like manner $P(N) = 0\cdot25$, $P(Q) = 0\cdot40$, and $P(U) = 0\cdot60$ are the marginal probabilities of selecting a non-industrial, a quoted, and an unquoted company respectively.

The figures in Table 15 have been carefully chosen, and exhibit a particular property, for

$$P(I) \times P(Q) = 0\cdot75 \times 0\cdot40 = 0\cdot30 = P(I, Q)$$
$$P(I) \times P(U) = 0\cdot75 \times 0\cdot60 = 0\cdot45 = P(I, U)$$
$$P(N) \times P(Q) = 0\cdot25 \times 0\cdot40 = 0\cdot10 = P(N, Q)$$
$$P(N) \times P(U) = 0\cdot25 \times 0\cdot60 = 0\cdot15 = P(N, U)$$

A check with the data of Table 14 reveals that this state of affairs does not always exist. For example there,

$$P(I) \times P(Q) = 0\cdot55 \times 0\cdot35 = 0\cdot1925 \neq 0\cdot25 = P(I, Q).$$

The special feature of the data used in this section is that the probability of a company belonging to one category is not influenced by its membership of the other category. Looking at this in more detail, the numbers of quoted companies that are industrial or non-industrial are in the ratio 3:1. Similarly, the numbers of unquoted companies that are industrial or non-industrial are in exactly the same ratio.

The two classifications in this particular case possess the property of *independence*. As a result of this independence, the joint probability of the two events I and U, say, is equal to the product of the marginal probabilities of I and U.

The discussion can be made quite general, and framed symbolically in the context of any two events A and B. If these events are statistically independent then

$$P(A, B) = P(A)P(B).$$

The property of independence can be alternatively expressed as

$$P(A \cap B) = P(A)P(B).$$

This follows from the identical interpretation which we can give to the joint probability $P(A, B)$ and the probability $P(A \cap B)$—i.e., each represents the probability that A and B both happen.

Worked Example

Verify, using the probabilities given to the elementary events in Chapter 2, Section 10, that increases in the prices of any two of the shares are independent.

Answer

Take shares 1 and 2. P(share 1 increases) = 1/3, P(share 2 increases) = 1/3, P(shares 1 and 2 both increase) = $P(L)$ = 1/9 (*see* Chapter 2, Section 11). Thus, P(shares 1 and 2 both increase) = P(share 1 increases) × P(share 2 increases).

A similar result obtains if any other pair from the three shares is considered.

Exercises on Section 2

1. Show in Chapter 2, Section 12, that J and L are not independent events.
2. Are two complementary events independent?
3. Statistical independence corresponds very well with the everyday use of the word "independence." Do you think that the following pairs of events are independent, or dependent?
    (*a*) Devaluation of the £.
       An increase in the Balance of Payments Deficit.
    (*b*) Increase in the Standard Rate of Tax.
       Decline in the sales of Winawag Dog Food.
    (*c*) Decline in the sales of Winawag Dog Food.
       Increase in the sales of Wuffalot Dog Food.
    (*d*) Decline in the sales of Winawag Dog Food.
       Decline in the sales of Wuffalot Dog Food.

4. In an attempt to increase sales revenue, the manager of the hosiery department in a store decides to test the effectiveness of a new style of advertising display. Over a period of 35 days he sometimes displays the advertisement, and sometimes does not.

His results are as follows:

	No. of days on which sales ⩾ £100	No. of days on which sales < £100
Advertisement used	10	15
Advertisement not used	3	7

Has he any reason to believe that his advertisement has a bearing on the level of sales? An error in the accounts subsequently reveals that sales exceeded £100 on one of the days when they were previously recorded as being less than £100. Is his conclusion unchanged (*a*) if this was a day on which the advertisement was not displayed? (*b*) if this was a day on which the advertisement was displayed? Comment on the reliability of his conclusions, bearing in mind that relative frequency is only an approximation to the true probability.

CONDITIONAL PROBABILITY

3. Introduction. As we shall see, it is only a short step from joint probability to *conditional* probability.

Consider the illustration from Chapter 2, Section 14, the data from which are reproduced in Table 16.

Table 16. Distribution of revenue from new product

Sample space of revenue (R) in first year (£000)	Probabilities of elementary events
$A \equiv \{R \geqslant 50\}$	0·20
$B \equiv \{50 > R \geqslant 40\}$	0·30
$C \equiv \{40 > R \geqslant 30\}$	0·20
$D \equiv \{30 > R \geqslant 20\}$	0·10
$E \equiv \{20 > R \geqslant 10\}$	0·10
$F \equiv \{10 > R \geqslant 0\}$	0·05
$G \equiv \{0 > R\}$	0·05

Suppose now that on the basis of this analysis the development is initiated, and that the director monitors its progress throughout the year. After 6 months he learns that revenue over the year will be definitely at least £20,000. The initial table of probabilities is no longer of direct use for he knows that E, F, G are now impossible events. How then can the director revise his analysis of revenue to be achieved by the year end?

He knows that events $J \equiv \{R \geqslant 20\}$ is now certain. Furthermore,

$$J = A \cup B \cup C \cup D$$

and A, B, C, D are mutually exclusive. If the probabilities of A, B, C, D, and J at this point in time are distinguished from those initially by writing $P(A \mid J)$, etc. (read "probability of A, given J has occurred" or "the conditional probability of A, given J") the statement

$$P(A \mid J) + P(B \mid J) + P(C \mid J) + P(D \mid J) = P(J \mid J) = 1$$

follows. Finally, since the original probabilities for A, B, C, and D were in the proportions 2:3:2:1, it seems reasonable to assign the probabilities, given J, in the same proportions to yield:

$$P(A \mid J) = \frac{2}{(2 + 3 + 2 + 1)} = 2/8$$

$$P(B \mid J) = \frac{3}{(2 + 3 + 2 + 1)} = 3/8$$

$$P(C \,|\, J) = \frac{2}{(2 + 3 + 2 + 1)} = 2/8$$

$$P(D \,|\, J) = \frac{1}{(2 + 3 + 2 + 1)} = 1/8$$

An alternative way to arrive at the same result is to argue that

$$\begin{aligned} P(J) &= P(A \cup B \cup C \cup D) \\ &= P(A) + P(B) + P(C) + P(D) \\ &= 0{\cdot}8 \end{aligned}$$

After 6 months have elapsed, the probability of J occurring has been increased by a factor of $1/P(J)$ from 0·8 to 1.

Increasing the probabilities of A, B, C, D by the same factor leads to

$$P(A \,|\, J) = P(A) \times \frac{1}{0{\cdot}8} = \frac{0{\cdot}20}{0{\cdot}80} = 2/8$$

$$P(B \,|\, J) = P(B) \times \frac{1}{0{\cdot}8} = \frac{0{\cdot}30}{0{\cdot}80} = 3/8$$

$$P(C \,|\, J) = P(C) \times \frac{1}{0{\cdot}8} = \frac{0{\cdot}20}{0{\cdot}80} = 2/8$$

$$P(D \,|\, J) = P(D) \times \frac{1}{0{\cdot}8} = \frac{0{\cdot}10}{0{\cdot}80} = 1/8$$

Exercise on Section 3

5. Calculate revised probabilities for the final revenue at the end of the year in the case where, 9 months after initiation of the development programme, revenue has already reached
   (*a*) £30,000;
   (*b*) £40,000.

4. Further development. It is true to say that all probabilities are conditional in the sense that they are influenced by the information available when they are calculated. However, the term "conditional" is usually restricted to cases where a revised set of probabilities is calculated in the light of some information which is additional to all the information that was available originally.

In the previous section the director's first probability values derived from his knowledge and experience of labour supply, economic factors, availability of raw materials, marketability of the product, etc. The second set are conditional probabilities calculated with the further knowledge that revenue had already reached £20,000.

In general, conditional probabilities are not obtainable in quite the simple way that has been discussed. Consider two depreciable assets *A* and *B* of the same type, and suppose that when they are purchased it is known that because of obsolescence and durability, their lives will not be more than 4 years in length. The sample space of possible outcomes may be broken down into elementary events representing the possible combinations of lives (in years), and probabilities may be assigned to each of them (Table 17).

Table 17. Probability distribution of service lives of two machines

Combination of Service Lives	Probability	Combination of Service Lives	Probability
1,1	0·09	1,3	0·05
2,1	0·07	2,3	0·07
3,1	0·03	3,3	0·10
4,1	0·01	4,3	0·08
1,2	0·07	1,4	0·03
2,2	0·12	2,4	0·04
3,2	0·05	3,4	0·05
4,2	0·03	4,4	0·11

When the assets are acquired the probability that, for example, the first will have a life of exactly 2 years and, in addition, the second of exactly 3 years, is assessed as 0·07. Notice that since the elementary events are exhaustive (it being assumed that lives are always of an integral number of years), and mutually exclusive, their probabilities total 1.

If at the end of 1 year it is found that the second asset has not survived, whilst an assessment of the first has not yet been made, the probabilities of the final possible outcomes can be updated. Firstly, the available information enables the sample space to be reduced to:

$$1,1 \quad 2,1 \quad 3,1 \quad 4,1.$$

Following the line of argument introduced in Section 3, these four elementary events form a mutually exclusive and exhaustive set.

The sum of their probabilities is unity, and apportioning this total probability in the same way as in the initial sample space (*i.e.*, 0·09 : 0·07 : 0·03 : 0·01 or 9 : 7 : 3 : 1) leads to revised probabilities of, respectively, 9/20, 7/20, 3/20, and 1/20. Denoting the life of the first asset by I, and of the second asset by II, this result can be restated in the language of conditional probabilities as:

$$P(\text{I} = 1 \,|\, \text{II} = 1) = 9/20$$
$$P(\text{I} = 2 \,|\, \text{II} = 1) = 7/20$$
$$P(\text{I} = 3 \,|\, \text{II} = 1) = 3/20$$
$$P(\text{I} = 4 \,|\, \text{II} = 1) = 1/20$$

Examining more closely the origin of these values note that, for example, 0·07 is the initial probability that both I = 2 and II = 1. In addition, 0·20 is the original (marginal) probability that II = 1 (since it is the sum of the probabilities of all the original elementary events involving II = 1). The conditional probability that I = 2, given that II = 1, can therefore be expressed in terms of probabilities in the initial sample space:

$$P(I = 2 \,|II = 1) = \frac{P\{(I = 2) \cap (II = 1)\}}{P(II = 1)}$$
$$= 0\cdot07/0\cdot20$$
$$= 7/20.$$

The remaining three conditional probabilities can be expressed in a similar way.

The results can now be generalised by direct extension of this argument. Suppose two events A and B are contained in a sample space, with probabilities of occurrence $P(A)$ and $P(B)$. The conditional probability $P(A\,|B)$ that A eventually occurs given that B has occurred already is calculated from:

$$P(A\,|B) = \frac{P(A \cap B)}{P(B)} \qquad (i)$$

It is evident that the conditional probability $P(A\,|B)$ is the ratio of the probability of the part of A that is in B, to the probability of B.

An interesting special case of equation (i) arises when event $A \cap B$ is impossible (*i.e.*, when A and B cannot both occur). Since $P(A \cap B)$ is then 0,

$$P(A\,|B) = \frac{P(A \cap B)}{P(B)} = \frac{0}{P(B)} = 0.$$

This result is expected, since if A and B are mutually exclusive events, A is impossible if B has already occurred. Finally, the equation might be thought to break down if $P(B) = 0$, since division by 0 is not possible. However, this condition labels B as an impossible event, and it is therefore meaningless to talk of $P(A\,|B)$.

5. Conditional probability and independence. We indicated in Section 2 above that in the general context of any two events A, B the events are independent if their joint probability, $P(A, B)$, is equal to the product of their individual probabilities, *i.e.*

$$P(A, B) = P(A) P(B)$$

A clear link between conditional probability and statistical independence can now be demonstrated, since

$$P(A \mid B) = \frac{P(A,B)}{P(B)} \qquad \text{by definition}$$

$$= \frac{P(A \cap B)}{P(B)} \qquad \text{from } (i) \text{ above}$$

$$= \frac{P(A) P(B)}{P(B)} \qquad \text{if } A, B \text{ are independent}$$

$$= P(A).$$

Interpreting this conclusion verbally, we can say that the probability of A happening, given that B has already happened, is equal to the unconditional probability of A happening. In other words, when A and B are independent the probability of A happening remains the same irrespective of whether B has already happened or not. Intuitively this is exactly what one would expect of two independent events.

Worked Example

In Chapter 2, Section 12 are events M and J independent or dependent? Justify your answer analytically, and verbally.

Answer

$J \equiv \{\text{share 3 does not increase in price}\}$,
$M \equiv \{\text{shares 1, 2, 3 behave similarly}\}$.

Examination of the sample space of Fig. 14 reveals that $J \cap M$ can occur in only two ways (*i.e.*, UUU, DDD). Thus, $P(J \cap M) = 2/27$. Similarly J' occurs in any of the ways: III, IDI, DII, UUI, UII, IUI, DDI, DUI, UDI. So $P(J') = 9/27$, and $P(J) = 1 - P(J') = 18/27 = 2/3$. Finally M can occur in any of the ways UUU, III, DDD, giving $P(M) = 3/27 = 1/9$.

Consequently, $P(J)P(M) = 2/3 \times 1/9 = 2/27 = P(J \cap M)$, and J and M are independent.

Alternatively, $P(J \mid M) = \dfrac{P(J \cap M)}{P(M)} = \dfrac{2/27}{1/9} = 2/3 = P(J)$.

Verbally, we can argue that the occurrence of M does not give us any information about the occurrence of J, since it does not indicate how share 3 will behave. Accordingly, the future occurrence of J is not affected by the occurrence of M.

Exercises on Section 5

6. In the example in Section 4 above calculate the following conditional probabilities:

(a) $P(\text{I} = 3 \mid \text{II} = 2)$ (b) $P(\text{II} = 4 \mid \text{I} = 2)$
(c) $P(\text{I} = 2 \mid \text{II} = 4)$ (d) $P(\text{II} = 1 \mid \text{I} = 2)$.

7. The behaviour of a stock quoted on the New York and London Stock Exchanges has been observed by a broker for some time. Over a 250-day period he has collected the following data on day-to-day changes:

	London Exchange Price			
	Unchanged	Decrease	Increase	
Unchanged	40	19	10	New York
Decrease	24	38	24	Exchange
Increase	27	20	48	Price

(On 40 days, for example, the price is unchanged on both exchanges.)

Explain carefully, using the notion of conditional probability, how the broker is able to use the fact that he obtains the London prices first to predict New York price behaviour.

8. In the example of Chapter 2 concerning the portfolio of 3 shares, find the probability that share price 3 has decreased.

Does the additional information about the behaviour of share 3 make any difference to the probability of A increasing?

9. Show in Chapter 2, Section 12, that J and L are not independent events.

10. Are two complementary events dependent or independent?

11. In Chapter 2, Exercises on Section 12, Question 17, given that a machine has lasted 3 years, what is the probability that it lasts:

   (a) just a further 2 years?
   (b) more than a further 2 years?

12. Rework Question 16, Exercises on Section 12, Chapter 2, in terms of conditional probabilities.

PERMUTATIONS AND COMBINATIONS

6. Permutations. The related topics of permutations and combinations will be required in later discussions of the binomial distribution and of non-parametric statistics. However, they are very useful in approaching basic problems involving probabilities, and are introduced at this stage to illustrate some of the ideas that have been developed in this and the previous chapter.

In the course of a sample audit let us suppose, for the sake of argument, that three accounts require examination. For identification purposes we will label these A, B, and C. It is not difficult to see that there are 6 orders in which the auditor may work his way through the accounts, *i.e. ABC, ACB, BAC, BCA, CAB, CBA*. Alternatively we say that there are 6 *permutations* of the 3 accounts.

Let us look more closely at how these 6 permutations arise. When the auditor is about to commence he has 3 choices open to him (*i.e.*, *A*, *B*, or *C*). Whichever he examines first, he then has two ways of choosing his second account (*e.g.* if he chose *A* first, he may choose either *B* or *C* second). Altogether, then, there are 6 ways in which he may choose the first two for scrutiny (*i.e.*, *AB*, *AC*, *BA*, *BC*, *CA*, *CB*). Finally, having examined two, there remains only one account for the final choice, giving a total of $3 \times 2 \times 1$ possible permutations.

This line of argument is readily extended to any number of accounts. Suppose, for example, there were five. To list all the permutations would be quite laborious, but in number there are

$$5 \times 4 \times 3 \times 2 \times 1 = 120.$$

In general the number of permutations of *n* accounts would be

$$n \times (n-1) \times (n-2) \times \ldots \times 3 \times 2 \times 1.$$

The following notation is used to avoid having to write the products in full,

$$3 \times 2 \times 1 = 3! \qquad \qquad \text{(read "factorial three")}$$
$$5 \times 4 \times 3 \times 2 \times 1 = 5!$$
$$n \times (n-1) \times \ldots \times 3 \times 2 \times 1 = n!$$

The number of permutations of a set of accounts (or, of course, of any other objects) increases very rapidly as the number of accounts increases. For example, $10! = 3,628,800$.

Taking our reasoning a little further, suppose that out of 5 accounts the auditor were required to examine only 3. How many ways could he do this? The answer is $5 \times 4 \times 3 = 60$, since for his first account he has 5 choices, for his second, 4, and for his third, 3. It is convenient to write this quantity in the form $5!/2! = 5!/(5-3)!$, since we may then make a general statement that if we have *n* objects, from which *r* are to be selected in order, the number of ways of doing so is $n!/(n-r)!$ This quantity is given the special symbol $_nP_r$ (read: the number of permutations of *n* objects, taken *r* at a time).

Thus,

$$_nP_r = \frac{n!}{(n-r)!}$$

As a particular case of this formula, the value $_nP_n$ is the number of permutations of *n* objects, *n* at a time, *i.e.* the number of permutations of all *n* objects. From the previous discussion, we know that

$$_nP_n = n!$$

and the formula, with $r = n$, gives

$$_nP_n = n!/(n-n)! = n!/0!$$

The only way in which the two expressions for $_nP_n$ can be reconciled is by setting $0! - 1$. The reader who is disturbed by this will find justification in the thought that there is only one way of ordering no objects!

7. Calculation of probabilities. We can regard the work of the auditor of Section 6 scrutinising his 5 accounts in terms of a random experiment. The sample space of the experiment contains all the permutations of the 5 accounts. We know that this sample space has 120 elements, representing the possible outcomes of the experiment. Assuming that the auditor makes his choice at random (so that no permutation is more likely to occur than any other), each outcome will have a probability of 1/120 occurring. Now suppose we are interested in the probability that account E is selected first together with account D second. All that we need to do is to find the number of elements in the sample space that begin with ED. Since to complete a permutation of 5 accounts beginning with ED, we require 3 members (A, B, C) which can themselves be permutated in $3 \times 2 \times 1 = 6$ ways, there must be 6 elements beginning with ED (*i.e.*, *EDABC*, *EDACB*, *EDBAC*, *EDBCA*, *EDCAB*, *EDCBA*). Each of these occurs with a probability of 1/120, so the probability of the event "auditor's examination begins with ED" is $6/120 = 1/20$.

As an illustration of the use of complementarity we next find the probability of the event "B is *not* chosen first." The number of elements in the sample space which begin with B is $4! = 24$, since there are just this number of permutations of the 4 remaining accounts.

Thus,

$$P(B \text{ is chosen first}) \quad = \frac{4!}{120} = \frac{1}{5}$$

$$P(B \text{ is } not \text{ chosen first}) = 1 - \frac{1}{5} = \frac{4}{5}.$$

Worked Example

In a ranking of 6 investment projects A, B, C, D, E, F calculate the probability that (*a*) B is first or C is second; (*b*) B is first, given that C is second.

Answer

Define events $L \equiv \{B \text{ is first}\}$,
$\qquad\qquad M \equiv \{C \text{ is second}\}$.
(*a*) Require $P(L \cup M)$,
$\qquad$ now $\quad P(L \cup M) = P(L) + P(M) - P(L \cap M)$.

Number of permutations of the 6 projects $= 6! = 720$.
Number of permutations with B first $= 5! = 120$.
Number of permutations with C second $= 5! = 120$.
Number of permutations with B first and C second $= 4! = 24$.

Thus,
$$P(L) = 120/720 = 1/6;$$
$$P(M) = 120/720 = 1/6;$$
$$P(L \cap M) = 24/720 = 1/30;$$

Finally,
$$P(L \cup M) = 1/6 + 1/6 - 1/30 = 3/10.$$
(b)
$$P(L|M) = P(L \cap M)/P(M)$$
$$= \frac{1/30}{1/6}$$
$$= 1/5.$$

(Notice that $P(L|M) \neq P(L)$, *i.e.* L and M are not independent events. The reason is the fact that if C is second (*i.e.*, M has "happened") the rankings that B can have are reduced from any of the 1st to 6th, to one of the 1st, 3rd, 4th, 5th, or 6th. The probability of L "happening" is therefore altered.)

Exercises on Section 7

13. A board of directors is composed of 12 men. In how many ways can the office of chairman, vice-chairman, and financial director be allocated assuming that one man can hold only one office?

14. There are three prizes in a local lottery. The first ticket in the draw wins the first prize, and so on. Altogether 600 tickets, numbered from 001 to 600 have been sold. In how many ways may the winning tickets be drawn? Mr. Avago, Mr. Better, and Mr. Chance hold respectively 3, 2, and 4 tickets. What is the probability that the first prize will go to Mr. Avago, the second to Mr. Better, and the third to Mr. Chance?

15. A company wishes to advertise three posts in "The Corporate Accountant." They are informed by the editors that an eight-page supplement in the next issue is to be devoted to vacancies. Each of their advertisements is to be placed on a different page. In how many ways may the advertisements be distributed through the supplement?

Before publication the company decides that it would prefer to have the posts advertised in order of seniority through the supplement. (The most senior job appearing on an earlier page than the second, and the junior post appearing later than the second.) In how many ways can the editors satisfy this requirement?

8. Combinations. The number of permutations, $_nP_r$, takes account not only of which r objects are selected, but also of their order of selection. In many situations, however, the order of selection is unimportant, and only the number of *combinations* is of interest.

As an example, in carrying out a sample audit by selecting 5 of the accounts, the composition of the sample is very relevant, whereas the order of selection of the members of the sample may well not matter.

Suppose a company has 5 projects *A, B, C, D, E* from which 3 are to be selected for investment. The

$$\frac{5!}{(5-3)!} = 60 \text{ permutations}$$

are listed in Table 18.

Table 18. Permutations of three projects from five

ABC	ABD	ABE	ACD	ACE
ACB	ADB	AEB	ADC	AEC
BAC	BAD	BAE	DAC	CAE
BCA	BDA	BEA	DCA	CEA
CAB	DAB	EAB	CDA	EAC
CBA	DBA	EBA	CAD	ECA
BCD	BCE	CDE	BDE	ADE
BDC	BEC	CED	BED	AED
CBD	CBE	ECD	EBD	EAD
CDB	CEB	EDC	EDB	EDA
DBC	EBC	DEC	DBE	DAE
DCB	ECB	DCE	DEB	DEA

Interest here is in combinations rather than permutations, and one can see that the 10 sub-groups into which Table 18 is divided give the 10 possible combinations of 3 projects from the 5:

ABC	*ABD*	*ABE*	*ACD*	*ACE*
BCD	*BCE*	*CDE*	*BDE*	*ADE*

Usually, however, it is not practical to list all the permutations of objects and regroup them into combinations. Denoting the number of combinations of 5 objects taken 3 at a time by $_5C_3$ we note that in the example,

$$_5C_3 \times 6 = {_5P_3}$$

Furthermore, the factor of 6 arises because in each combination of three projects, there are $3! = 6$ permutations. Consequently the number of combinations, times the number of permutations within each combination, gives the total number of permutations. Thus, in general,

$$_nC_r \times r! = {_nP_r}$$

or

$$_nC_r = \frac{_nP_r}{r!}$$

$$= \frac{n!}{(n-r)!\,r!}$$

An interesting insight into the nature of combinations can be gained by noting that

$$_5C_3 = \frac{5!}{3!\,2!} = {}_5C_2$$

$$_8C_2 = \frac{8!}{2!\,6!} = {}_8C_6$$

and generally,

$$_nC_r = \frac{n!}{(n-r)!\,r!} = {}_nC_{n-r}$$

The reason for this relationship lies in the fact that every combination of r objects from n is automatically accompanied by a combination of $(n-r)$ objects, *viz.* those not selected. Consequently $_nC_r$ and $_nC_{n-r}$ are equal. As an illustration, there is an exact correspondence between the 10 combinations of any 3 of the projects, and 10 of remaining 2 projects:

ABC	*DE*
BCD	*AE*
ABD	*CE*
BCE	*AD*
ABE	*CD*
CDE	*AB*
ACD	*BE*
BDE	*AC*
ACE	*BD*
ADE	*BC*

A final point concerns $_nC_n$, which is the number of combinations which include all n objects.

The answer is clearly one only, and (using $0! = 1$) the formula for $_nC_n$ is in agreement with this,

$$_nC_n = \frac{n!}{(n-n)!\,n!}$$

$$= \frac{n!}{0!\,n!}$$

$$= 1$$

9. Calculation of probabilities using combinations. A probabilistic element is introduced into combinations in precisely the manner of Section 7 above. The procedure is again based on first determining the size of the sample space in a particular situation (*i.e.*, finding the total number, T, of possible combinations). The probability of a particular event can then be found by obtaining the number of elements, E, in the sample space which are contained in the event. Assuming all the elements in the sample space are equally likely the required result is E/T. The following two examples clearly illustrate the calculations involved.

Worked Examples

1. In Section 8 above, find (*a*) the probability that A and B will both be chosen, (*b*) the probability that A has been chosen, given that news has leaked out that E has not been chosen.

Answer

(*a*) Examining the list of 10 possible combinations (each of which will have a probability of 1/10 attached to it, since selection is random) we note that there are 3 which include both A and B. The required probability is therefore 3/10.

(*b*) The number of combinations which include A, but exclude E, is the number of ways of choosing two projects from BCD to supplement the presence of A in the final selection, *i.e.* $_3C_2 = 3$.

Thus,
$$P(A \text{ chosen, and } E \text{ not chosen}) = 3/10.$$

Also the number of combinations which exclude E is the number of ways 3 projects can be selected from $ABCD$, *i.e.* $_4C_3 = 4$.

Thus,
$$P(E \text{ not chosen}) = 4/10.$$

Finally, the required probability is

$$P(A \text{ chosen} \,|\, E \text{ not chosen}) = \frac{P(A \text{ chosen, and } E \text{ not chosen})}{P(E \text{ not chosen})}$$
$$= \frac{3/10}{4/10}$$
$$= 3/4.$$

2. A company has recently changed to a computerised system of accounts. It is decided to monitor the performance of the system closely over the first few weeks. The total set of accounts is subdivided into 7 different categories, and it is proposed that a category should be chosen at random each Monday, Wednesday, and Friday for close scrutiny. Repetitions are not excluded. Find

(*a*) The number of possible patterns of investigation in a week.
(*b*) The number of possible combinations of checks in a week.
(*c*) The probability that on just 2 of the 3 days the same category will be checked.
(*d*) The probability of any repetition occurring.
(*e*) The probability that category A is checked more than once.
(*f*) The probability that A was checked, given that E was checked.

Answer

(*a*) Repetitions are not excluded, so that each week the number of possible patterns of investigation is not $_7P_3$, but $7 \times 7 \times 7 = 343$ (since each day there are 7 ways of selecting a category).

(*b*) Denoting the categories of account by *A, B, C, D, E, F, G*, the possible combinations may themselves be broken down into different types:

(*i*) Different on all 3 days (*e.g. ACB* or *DBA*).
(*ii*) Same on 2 days, different on third day (*e.g. ACA* or *DDB*).
(*iii*) Same on all 3 days (*e.g. AAA* or *BBB*).

A few moments' thought should convince the reader that (*i*), (*ii*), (*iii*) are exhaustive and mutually exclusive sets. Turning now to the number of combinations of each type there are $_7C_3 = 35$ in (*i*). In (*ii*) there are 7 ways in which the like pair can occur, together with 6 ways for the remaining category to be chosen, giving a total of $7 \times 6 = 42$ combinations of this type. Finally, there are 7 combinations of type (*iii*).

The total number, *T*, of possible combinations of checks in a week is therefore:

$$T = 35 + 42 + 7$$
$$= 84.$$

(*c*) Since selection is made at random each day, each and every combination occurs with a probability of 1/84. The probability, therefore, that on just 2 of the 3 days the same category will be checked is $42/84 = 0.50$.

(*d*) Since we require the probability of any repetition at all, category (*iii*) must be also included to give a probability of $49/84 = 0.58$.

(*e*) The probability that a specific category, say *A*, is checked more than once is $7/84 = 0.083$ since there are 6 combinations of type (*ii*) which include two *A*s and one of type (*iii*).

(*f*) $$P(A \text{ was checked} \,|\, E \text{ was checked})$$

$$= \frac{P(A \text{ and } E \text{ both checked})}{P(E \text{ was checked})}.$$

Now *A, E* are both checked in 5 combinations of type (*i*), in 2 combinations of type (*ii*) and in no combinations of type (*iii*), thus

$$P(A \text{ and } E \text{ both checked}) = 7/84.$$

Also, *E* is checked in $_6C_2 = 15$ combinations of type (*i*) (since there are $_6C_2$ ways in which two different categories can be selected from the remaining 6 to supplement *E*), in 12 combinations of type (*ii*) (*i.e. EEA, . . ., EEG, EAA, . . ., EGG*), and in one combination of type (*iii*).

Thus,

$$P(E \text{ was checked}) = \frac{15 + 12 + 1}{84} = 28/84.$$

Consequently,

$$P(A \text{ was checked} \,|\, E \text{ was checked}) = \frac{7/84}{28/84} = 1/4.$$

Exercises on Section 9

Questions 16–19 all relate to the worked example concerning the computerised accounting system in Section 9 above.

16. If a different category is to be considered on each of the 3 days per week, what is the probability that categories *A* and *B* are both checked?

17. Under the same conditions as Question 16, what is the probability that:

(*a*) *A* is not checked?
(*b*) neither *A* nor *B* is checked?
(*c*) *A* is checked given that *D* is not?

18. If repetitive checking is allowed, what are the probabilities of (*a*), (*b*), and (*c*) in Question 17?

19. Owing to a national holiday on Monday, suppose that checks are only performed on Wednesday and Friday in a particular week. If repetitive checkings are allowed what is the probability that:

(*a*) *A* is checked, but *D* is not?
(*b*) *A* and *B* are both checked?
(*c*) neither *A* nor *B* is checked?

20. A computerised accounting system involves punched card storage of data on accounts—one card for each account. Each card has 60 columns in which any digit 0–9, or any letter other than I or O, can be punched. The first 3 columns are used to identify the account. How many accounts may be identified if letters and numbers can both be used, but no account can contain any repetitions in its identification?

If all possible identities have been used, what is the probability that a randomly selected account will contain a 2 in its identity number? Given that the account drawn has a 2 in it, what is the probability that it has also a 1 or an A in it?

21. Blott Inc. has 3 possible independent investment projects in mind. The internal rate of return on each of the 3 projects can assume any integer value between 8 and 15 per cent inclusively. The rates of return have equal probabilities. What is the probability

(*a*) that at least 2 of the projects will realise rates of return in excess of 10%?
(*b*) that all 3 will realise returns less than 12%?

22. An investigation is undertaken into the profitability of 48 companies. For the purposes of the survey the companies are divided into 4 groups— each with 12 members—according to whether they are in food, manufacturing, mining, or are stores. One company is selected from the 48 and studied in depth. The second company selected is in the manufacturing sector. What is the probability that the first one was also in that sector?

23. In Exercises to Section 7, Question 13 above, a committee of the board is set up to consider a particular problem in the company.

In how many ways can this be done, assuming that the Chairman is, *ex officio*, a member of the committee?

After some deliberation the committee decides to co-opt two further members. In how many ways can this be done?

CHAPTER 4

PROPERTIES OF PROBABILITY
DISTRIBUTIONS

INTRODUCTION

1. Random variables. In the vast majority of statistical work one is concerned with sample spaces and events that are of a numerical nature. Granted that probabilities can be assigned to events like "it will rain tomorrow" or "that machine will break down next time it is operated," events of the type "profits will be £6000," "that machine will survive 4 years," and "that company will issue £40,000 worth of loan stock" are much more common. Because of the importance of a variable (such as profit, life, worth) which can take with uncertainty one of a number of different values, it is given a special name— *a random variable*. Examples of random variables are:

(*a*) The number of times that the bank rate changes in 1985–1990.
(*b*) The value of a company's sales in 1982.
(*c*) The number of members of the I.C.A.E.W. next year.

On the other hand:

(*a*) The number of pages in this book,
(*b*) The current standard rate of tax,
(*c*) A company's declared profits in 1972,

are not random variables, since each of the values is known with certainty.

The values that a random variable can assume are usefully considered as elementary events in a sample space which contains all these possible values. For convenience, random variables are placed into one of two main categories—the *discrete* and the *continuous*. The former category contains random variables that assume only isolated numerical values. For example, the number of members of the I.C.A.E.W. next year can only take one of the integer values 0, 1, 2, 3, . . ., and not, say, 2000½. (Here the sample space is made up of the elementary events "the random variable assumes value 0," "the random variable assumes value 1," etc., or more concisely, 0, 1, etc.) Similarly the percentage of payments due out of a total of 1000 that will still be outstanding at the end of the year can only be 10·3%,

10·4% etc., and not, say, 10·35%. This second example illustrates the point that a discrete random variable is not necessarily restricted to just integer values. Graphically, the characteristic feature of a discrete random variable is that it takes single values on a scale from $-\infty$ to $+\infty$ thus:

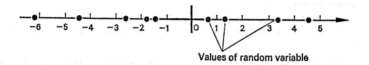

Values of random variable

The number of members of the I.C.A.E.W. is limited in theory by the number of people on earth, and the percentage of outstanding payments can only take one of a limited number of values.

However, it is not invariably the case that an upper limit can be placed on the number of elementary events in the sample space of values of a discrete (or, for that matter, a continuous) random variable. Thus, an actuary interested in the number of motor accidents next year would find it impossible to say that there cannot be more than say 10,000 or 20,000. For all intents and purposes the sample space in this case contains an infinite number of elementary events, *viz.* 0, 1, 2, 3, ...

A continuous random variable is rather different in nature from the discrete type. As its name suggests the values it can take form a continuous range. Take the case of the insurance company again, and suppose that it is interested in the exact ages that its current life policy-holders will attain. Clearly, a person's age at death can in theory be determined to any degree of accuracy—years, hours, minutes, seconds, and fractions of a second. A person's age at death can therefore take any value in a continuous range from 0 to 200 years. (The upper limit here has been chosen unrealistically high since there are claims of longevity of the order of 180 years!) The random variable "age at death" has an infinite number of elementary events in its sample space, since one could not list all the possibilities. A moment's thought should convince the reader that this is true of continuous random variables in general, since the very fact that it is continuous implies that between any two specified values another can always be found.

In practice, of course, the insurance company cited would, for convenience, divide the sample space of possible ages into blocks of one year and treat the random variable as a discrete one. Conversely, the net present value of a project in a capital budgeting situation might be treated as a continuous random variable, although strictly speaking it is discrete with gaps of $\frac{1}{2}$p between the possible values.

Exercises on Section 1

1. State in each of the following cases whether the random variable is discrete or continuous:

    (*a*) Market value of a share tomorrow.

    (*b*) Declared profits of a company next year.

    (*c*) Quantity of liquid fertiliser produced by a chemical company in the coming week.

    (*d*) Time taken to attend to a client.

    (*e*) Quantity of copper ore mined by a company in the next financial year.

2. Define suitable sample spaces for the random variables (*a*)–(*e*) in Question 1.

2. Discrete probability distributions.

The real usefulness of a random variable lies in the association of probabilities with each of its possible values. Since one of the possible values must occur, a total probability of 1 is available for distribution amongst these values. A complete specification of this breakdown is referred to as the *probability distribution of the random variable*. The distribution is discrete or continuous, depending on the type of random variable with which it is associated.

Suppose that the future cost of capital is assessed by a company to be either 10%, 11%, 12%, 13%, or 14%. The random variable "cost of capital" will assume one of these five values. The financial director might consider that the associated probability distribution takes the form:

Random Variable (Cost of Capital) (%)	Probability distribution
10	0·20
11	0·40
12	0·30
13	0·05
14	0·05

The elementary events are, of course, mutually exclusive and exhaustive, resulting in a set of probabilities with a total of 1. Since the random variable has been defined in a discrete way the probability distribution is also discrete.

A random variable is often denoted by an upper case letter, in this case perhaps C (for cost), with the corresponding lower case letter representing a value that it can take. This allows of a very compact statement of the probability properties of the random variable. Thus $P(C = c)$ reads "the probability that random variable C has the value c." In the preceding example

$$P(C = 10\%) = 0·20, \ P(C = 11\%) = 0·40.$$

This notation is particularly convenient in connection with the more commonly occurring probability distributions such as the binomial distribution, which will be dealt with in more detail in Chapter 5. If the prices of n shares in a portfolio vary independently, and each has a probability of increasing of 0·5, the number of shares which do actually increase is a discrete random variable, X say, with possible values $x = 0, 1, 2, \ldots n$. It can be shown that

$$P(X = x) = \frac{n!}{(n - x)!\, x!} \left(\frac{1}{2}\right)^n$$

Here the probability distribution of X has been completely specified in one single formula.

To find the probability that X assumes any particular one of its possible values, it is necessary to substitute that value in the formula.

Thus:

$$P(X = 3) = \frac{n!}{(n - 3)!\, 3!} \left(\frac{1}{2}\right)^n$$

If $n = 5$ then,

$$P(X = 3) = \frac{5!}{2!\, 3!} \left(\frac{1}{2}\right)^5$$
$$= 10/32$$
$$= 0\cdot3125.$$

This, therefore, is the probability that 3 shares in a 5-share portfolio will increase in price under the stated conditions.

This notation can be extended simply to cover situations where the probability of a random variable assuming a value in a range is required.

Still with $n = 5$, the probability that not less than 2 and not more than 4 shares increase in price can be written concisely as:

$$P(2 \leqslant x \leqslant 4)$$
$$= P(x = 2) + P(x = 3) + P(x = 4)$$
$$= 0\cdot3125 + 0\cdot3125 + 0\cdot1563$$
$$= 0\cdot7813.$$

Exercises on Section 2

3. In a sample of n accounts the probability that the number of delinquent accounts is r is given by:

$$P(R = r) = \frac{n!}{(n - r)!\, r!} (0\cdot1)^r (0\cdot9)^{n-r}$$

(This is a further example of a binomial distribution.)

If $n = 6$, find

(a) $P(R = 5)$,
(b) $P(R \leqslant 3)$.

(HINT: calculate $P(R = 0) + P(R = 1) + P(R = 2) + P(R = 3)$.)

(c) $P(R > 3)$.
(d) $P(R \leqslant 5)$,
(e) $P(1 \leqslant R \leqslant 4)$.

3. Continuous probability distributions.

Continuous probability distributions, associated with continuous random variables, require rather more careful definition that in the corresponding discrete case. The difficulty arises in attempts to allocate probabilities to an infinite number of values of the random variable in such a way that the probabilities total 1.

Suppose an electric clock has a sweep second-hand (which moves smoothly around the face of the clock). The position in hours of the hand on the face is a continuous random variable taking any value between 0 and 12. If the clock is observed at random, what is the probability that the second hand is in exactly the 3 o'clock position?

The motion of the second-hand and the manner of observation lead one to believe that no one position is any more likely to occur than any other. Unfortunately, this perfectly reasonable assumption leads to a seeming impasse. By dividing the interval between two consecutive hours on the face into 100 parts, 1200 possible positions for the hand can be defined. Admittedly there will be only a small gap between each of these positions, but this does not detract in any way from their separate identities. Assigning each of these 1200 equi-probable positions a probability of, say, 1/1000 leads to a total probability in excess of 1, *viz.* 1·2. To attempt to overcome this objection, a smaller probability, say 1/10,000, could be assigned to each position. Again, however, since 20,000 positions of the continuously moving second-hand can be defined, one is still faced with the prospect of an inadmissible total probability greater than 1.

It is not difficult to see now that, however small a probability is assigned to each position of the hand, since the number of positions is infinite, the total probability will exceed 1. The only way out of the dilemma is to attach zero probability to each position. Certainly the second-hand can occupy the exact position "3 o'clock," say, since it passes through that position every time it goes from 2 to 4. However, the probability that it is in the exact position when the clock is observed is infinitesimally small, since there are so many other positions it could occupy.

Although the assignation of zero probabilities to exact values may seem logical, it does not at first sight seem very useful. If, however, instead of attempting to find the probability of particular positions occurring, we look at *ranges* of positions, the whole matter is resolved. For example, nothing could be more natural than to say that the probability of the hand lying between 6 and 7 is 1/12, since this range is 1/12 of the total range around the face. Similarly, the probability that the hand lies between 4 and 9 is 5/12.*

The results of the preceding argument are summarised graphically in Fig. 21.

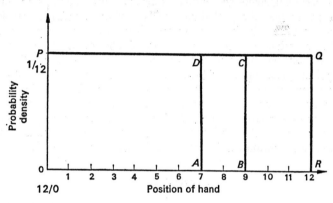

FIG. 21.—Continuous probability distribution of position of sweep second hand on clock face.

Rectangle OPQR is 12 units long and 1/12 of a unit high, so that its area is one square unit. This area represents the total probability of one. The area standing on any interval and bounded by PQ represents the probability of the hand being in that particular interval. To find the probability that the hand lies in the range 7–9 one erects lines AD, BC through 7 and 9 respectively. This defines rectangle ABCD which is of height 1/12 and base 2. Its area, 1/6, is the probability attached to range 7–9.

*It is interesting to note that it is immaterial whether the end-points of the range are included or not. For instance,

$$P(4 \leqslant \text{position of hand} \leqslant 9) = P(\text{position of hand} = 4)$$
$$+ P(4 < \text{position of hand} < 9)$$
$$+ P(\text{position of hand} = 9)$$

since the three events on the right are mutually exclusive and exhaustive subsets of the event on the left. Now,

$$P(\text{position of hand} = 4) = P(\text{position of hand} = 9) = 0$$

So,

$$P(4 \leqslant \text{position of hand} \leqslant 9) = P(4 < \text{position of hand} < 9).$$

This example is of a particularly simple type of continuous distribution known as the *uniform distribution*—since the total probability of one is evenly distributed over the range of the random variable. The height of the figure whose area represents probability is referred to as the *probability density* of the random variable. In the case of the uniform distribution the probability density has a constant value $1/n$, where n is the length of the range of the random variable.

4. General comments on continuous distributions. There is no reason to suppose that the probability density of any random variable is always constant, in fact the uniform distribution is one of the least useful distributions. Other more useful and common continuous distributions will be examined in more detail in Chapter 6. All have a graphical presentation, which includes an explicit mathematical expression for the height y of the curve (*i.e.*, probability density) at any place, in terms of the value x of the random variable there. This mathematical relationship is called the *probability density function* of the random variable. All the relevant information about any continuous distribution can be expressed in the form of Fig. 22.

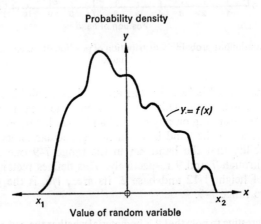

FIG. 22.—General continuous probability density function.

Here x_1 and x_2 mark the ends of the range of possible values of the random variable (either or both of these may be infinite). The probability density function, $y = f(x)$, defines the boundary of the area representing the total probability of one.

Thus,

$$\int_{x_1}^{x_2} f(x)dx = 1.$$

Rather more generally, the probability that the random variable, X, assumes a value x in the range x', x'' is given by the area underneath the curve $y = f(x)$ between x' and x'', i.e.

$$P(x' < x < x'') = \int_{x'}^{x''} f(x)dx.$$

By way of illustration, the probability density function $y = e^{-x}$ ($0 < x < \infty$) may represent the distribution of life in years (and parts of a year) of a depreciable asset (*see* Fig. 23).

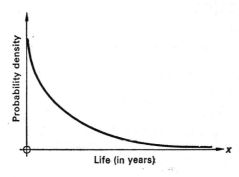

FIG. 23.—Distribution of life of an asset (exponential distribution).

Note that,

$$\int_0^{\infty} e^{-x} \, dx = \left[-e^{-x} \right]_0^{\infty}$$
$$= [0] - [-1]$$
$$= 1$$

The total area underneath the curve is unity, giving a total probability of 1. This verifies that the function $y = e^{-x}$ is a legitimate probability density function. Note also that according to this (theoretical) model of depreciation no upper limit has been fixed for the life of the asset.

To find the probability that the life is of between $\frac{1}{2}$ and $1\frac{1}{2}$ years one proceeds as follows:

$$P(\tfrac{1}{2} \leqslant x \leqslant 1\tfrac{1}{2}) = \int_{\frac{1}{2}}^{\frac{3}{2}} e^{-x} \, dx$$
$$= \left[-e^{-x} \right]_{\frac{1}{2}}^{\frac{3}{2}}$$
$$= [-e^{-\frac{3}{2}}] - [-e^{-\frac{1}{2}}]$$
$$= e^{-\frac{1}{2}} - e^{-\frac{3}{2}}$$
$$= 0 \cdot 6065 - 0 \cdot 2231$$
$$= 0 \cdot 3834.$$

Exercises on Section 4

4. The internal rate of return, $x\%$, of a capital investment project has a probability density function (p.d.f.)

$$\frac{2}{9}(x - 9)(12 - x).$$

Sketch the graph of this function, and show that it is a legitimate p.d.f. Why is it impossible for the rate of return to be less than 9% or greater than 12%? What is the probability that the internal rate of return is

    (*a*) between 10% and 11%?
    (*b*) less than 10%?
    (*c*) less than 11%?

5. In the example at the end of Section 4 calculate the probability that the life is

    (*a*) not more than 1 year;
    (*b*) more than 1 year;
    (*c*) between 3 and 9 months;
    (*d*) less than 6 months.

NOTE: Those unable to perform the necessary integrations may find approximate values for the areas by plotting the curves on graph paper and counting the small squares under the curves.

5. The relative frequency approach to probability.

Table 19 gives industrial mortality figures as used by a company employing a probabilistic method of depreciation.

Table 19. Distribution of asset lives

Age at failure, A (*in months*)	Number of machines	Proportion of machines
$0 \leqslant A < 12$	6	0·03
$12 \leqslant A < 24$	6	0·03
$24 \leqslant A < 36$	10	0·05
$36 \leqslant A < 48$	12	0·06
$48 \leqslant A < 60$	20	0·10
$60 \leqslant A < 72$	30	0·15
$72 \leqslant A < 84$	38	0·19
$84 \leqslant A < 96$	44	0·22
$96 \leqslant A < 108$	24	0·12
$108 \leqslant A < 120$	10	0·05
Total	200	1·00

The relative frequency histogram of the data has the form shown in Fig. 24.

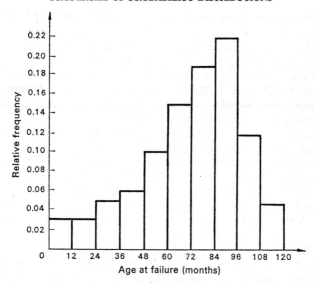

FIG. 24.—Relative frequency histogram for data of Table 19.

Suppose now that instead of using a class interval of 12 months an interval of 1 month was taken, and data was collected on 2400 machines, rather than 200. The resulting relative frequency histogram might look like Fig. 25.

Again (in theory) the class interval could be shortened to 1 week,

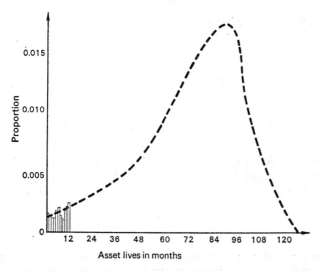

FIG. 25.—Theoretical probability distribution of asset lives.

and the number of machines increased to 9600, giving a histogram with an even smoother profile than the last. *In principle* one can continue this procedure indefinitely, since age is a continuous variable. At each stage, the sample of machines would be increased in proportion to the reduction in class interval, otherwise the number of class intervals would soon exceed the number of machines. Eventually a point would be reached at which the profile of the histogram would be indistinguishable from a smooth curve.

The area of each constituent rectangle represents a proportion of the total area, implying that the area underneath this smooth curve will be unity. Since relative frequency has already been introduced as an estimate of "true" probability, it is only a small step to suggest that the true probability distribution of ages is approximated by a relative frequency histogram. The larger the total number of machines, the better this approximation will be. The theoretical limiting smooth curve can now be recognised as the probability density of the random variable, "age at failure."

Although much accounting and financial data is, strictly speaking, of a discrete nature, the value of the continuous type of distribution lies in its relative ease of handling in the mathematical sense, and in the fact that it represents a theoretical "ideal" distribution to which a particular set of data approximates.

MEAN AND VARIANCE OF PROBABILITY DISTRIBUTIONS

6. Introduction. The summary statistics measuring central tendency and dispersion have been introduced in the context of frequency distributions. Because of the close link between relative frequency and probability, and indeed the fact that the two statistics measure properties of distributions, it is not surprising to learn that analogous concepts are used in relation to probability distributions. In fact these measures are the first two of a series of quantities known as the *moments* of the (frequency or probability) distribution.

7. Expected value of a probability distribution. In the case of a discrete frequency distribution, if x_i is a datum value, f_i its frequency of occurrence, and n the total number of measurements, the mean is defined as:

$$\sum_{i=1}^{n} \frac{f_i x_i}{n} = \sum_{i=1}^{n} \frac{(f_i)}{n} x_i$$

Now f_i/n is the relative frequency of occurrence of the value x_i. Accordingly f_i/n approximates to the true probability p_i of x_i occurring. Immediately this suggests that a suitable measure of centrality for the related probability distribution might be:

$$\sum_{i=1}^{n} p_i x_i$$

This quantity is called the *expected value*, or *expectation*, of the discrete random variable (which takes values x_i with probabilities p_i). Rather loosely it is sometimes called the mean, although this term is better restricted to use with frequency distributions.

By way of illustration, in Section 2 above, the random variable cost of capital, and its probability distribution, were given:

Values of Random Variable (%)	Probability
10	0·20
11	0·40
12	0·30
13	0·05
14	0·05

The expected cost of capital has the value

$$(0·20 \times 10\%) + (0·40 \times 11\%) + (0·30 \times 12\%) \\ + (0·05 \times 13\%) + (0·05 \times 14\%) = 11·35\%.$$

For a continuous probability distribution the concept of expected value is unchanged, although the methods of calculation in the discrete and continuous cases appear to differ.

The argument depends on a knowledge of elementary calculus, and it is unnecessary to go into this in detail. In fact the continuous analogue to the expected value of a discrete distribution is given by:

$$\int_{x_1}^{x_2} xp(x)dx$$

where $p(x)$ is the probability density function, and x_1, x_2 are the upper and lower limits of the random variable.

The notation $E(x)$ is used for the expected value of a (discrete or continuous) random variable, X.

As an example of this calculation the more mathematically practised reader may be interested in the following. Consider the situation suggested in Exercises on Section 4 above, Question 4. The graph of the probability density function is shown in Fig. 26.

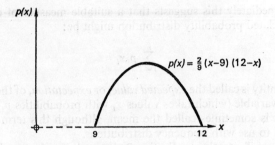

$$p(x) = \tfrac{2}{9}(x-9)(12-x)$$

FIG. 26.—Graph of probability density function $p(x) = \tfrac{2}{9}(x - 9)(12 - x)$.

Expected value of internal rate of return

$$= \int_9^{12} \frac{2}{9} x \, (x - 9) \, (12 - x) \, dx$$

$$= \frac{2}{9} \int_9^{12} (-x^3 + 21x^2 - 108x) \, dx$$

$$= \frac{2}{9}\left[-\frac{x^4}{4} + 7x^3 - 54x^2 \right]_9^{12}$$

$$= \frac{2}{9}(-5184 + 12{,}096 - 7776) - \left(\frac{-6561}{4} + 5103 - 4374 \right)$$

$$= \frac{2}{9}(-864 + 911\cdot25)$$

$$= \frac{2}{9} \times 47\cdot25$$

$$= 10\cdot5.$$

Since the graph of the density function is symmetrical and lies between 9 and 12, one could anticipate this result.

8. Variance of a probability distribution. Returning once more to a discrete frequency distribution, and introducing the notation $\bar{x}$ for the mean, the variance is given by:

$$\sum_i f_i(x_i - \bar{x})^2/n = \sum_i (f_i/n)(x_i - \bar{x})^2.$$

For the corresponding discrete probability distribution f_i/n and $\bar{x}$ are replaced by the true values of probability (p_i) and expectation (μ) which they respectively estimate. As a result the variance of a discrete random variable, X, assuming values x_i with probabilities p_i is defined by:

$$\text{var } X = \sum p_i(x_i - \mu)^2.$$

Using the cost of capital example:

Variance of cost of capital

$$= 0.20 (10 - 11.35)^2 + 0.40 (11 - 11.35)^2 + 0.30 (12 - 11.35)^2$$
$$+ 0.05 (13 - 11.35)^2 + 0.05 (14 - 11.35)^2$$
$$= (0.20 \times 1.8225) + (0.40 \times 0.1225) + (0.30 \times 0.4225)$$
$$+ (0.05 \times 2.7225) + (0.05 \times 7.0225)$$
$$= 1.0275,$$

and standard deviation $= \sqrt{(1.0275)}$

$$= 1.014.$$

The analogous definition of the variance for a continuous probability distribution is:

$$\text{var } X = \int_{x_1}^{x_2} (x - \mu)^2 \, p(x) dx.^*$$

The reader need not concern himself too much with the actual evaluation of these integrals, since these can be tedious. In reality much of our work will be based on a small number of distributions whose variances (and means) are well-known.

Worked Examples

1. Calculate the mean and variance of the uniform distribution with density function $p(X) = 1/n$ over the range $0 \leqslant x \leqslant n$.

Answer

(a) $\text{E}(x) = \int_0^n x \cdot 1/n \, dx = \frac{1}{n}\left[\frac{x^2}{2}\right]_0^n = \left(\frac{n^2}{2} - 0\right)\Big/ n = \frac{1}{2}n$

(b) $\text{var } X = \int_0^n \left(x - \frac{n}{2}\right)^2 \frac{1}{n} \, dx = \frac{1}{n}\int_0^n \left(x^2 - nx + \frac{n^2}{4}\right) dx$

$$= \frac{1}{n}\left[\frac{x^3}{3} - \frac{nx^2}{2} + \frac{n^2 x}{4}\right]_0^n = \frac{1}{n}\left[\frac{n^3}{3} - \frac{n^3}{2} + \frac{n^3}{4}\right]$$

$$= n^2/12.$$

* From the preceding definition of the expected value of a continuous random variable, we note that by treating the quantity $(X - \mu)^2$ as a random variable itself, the definition of var X can also be written as

$$\text{var } X = \text{E}(X - \mu)^2.$$

2. A news-stand operator assigns probabilities to the weekly demand for *Roué* magazine as follows:

Demand (copies)	Probability of this demand
10	0·10
11	0·15
12	0·30
13	0·25
14	0·20
Total	1·00

An issue sells for 50p and costs him 30p.

(*a*) If the operator can return free of charge any unsold copies, how many should he order?

(*b*) If the operator cannot return unsold copies, how many copies should he order?

HINT: Use the criterion of expected value of sales.

Answer

(*a*) At least 14 copies, in order that no sales opportunities are lost. In fact it is immaterial to him how many more than 14 are ordered, since there is no penalty for overstocking.

(*b*) There would be a penalty incurred for certain if more than 14 were ordered, so he should place an order for 14 or less:

Orders 10 copies:
Expenditure = £3·00
Expected income = £5·00
Expected net profit = £2·00

(in fact this profit is certain since demand is always at least 10).

Orders 11 copies:
Expenditure = £3·30
Expected income = £{(5·00 × 0·10) + (5·50 × 0·90)} = £5·45
Expected net profit = £2·15

(note that the probability of an income of £5·50 is not 0·15, since if demand is 12, 13, or 14 income will still be £5·50).

Orders 12 copies:
Expenditure = £3·60
Expected income = £{(5·00 × 0·10) + (5·50 × 0·15)
$$+ (6·00 × 0·75)\} = £5·82\tfrac{1}{2}$$
Expected net profit = £2·22½.

Orders 13 copies:
Expenditure = £3·90
Expected income = £{(5·00 × 0·10) + (5·50 × 0·15)
$$+ (6·00 × 0·30) + (6·50 × 0·45)\} = £6·05$$
Expected net profit = £2·15.

Orders 14 copies:
  Expenditure — £4·20
  Expected income = £{(5·00 × 0·10) + (5·50 × 0·15)
                          + (6·00 × 0·30) + (6·50 × 0·25)
                          + (7·00 × 0·20)} = £6·15
  Expected net profit = £1·95.

By ordering 12 copies per week the operator will maximise his net profit on average (assuming that the pattern of demand does not change).

Exercises on Section 8

6. The distribution of individual incomes exceeding a lower limit x_0 sometimes has the form known as a *Pareto distribution*. The probability density function is

$$\frac{k}{x_0}\left[\frac{x_0}{x}\right]^{k+1} \quad (x > x_0, k > 1)$$

Sketch the curve of this function, and find the expected value and variance of income when $x_0 = £1000$ and $k = 2$.

7. Two shares, Bits and Bats, both have a current price of £2·30. The closing prices of the two shares at the end of the week are estimated to have the probability distributions:

Bits		Bats	
Price	*Probability*	*Price*	*Probability*
2·20	0·125	2·20	0·04
2·25	0·125	2·25	0·08
2·30	0·125	2·30	0·12
2·35	0·125	2·35	0·30
2·40	0·125	2·40	0·18
2·45	0·125	2·45	0·16
2·50	0·125	2·50	0·08
2·55	0·125	2·55	0·04

If you intend to buy one of the shares now, to sell at the end of the week, on the basis of expected value and variance, which would you choose and why?
      HINT: Investment analysts sometimes use standard deviation as a surrogate for risk.

8. Penny Wise lives in the women's hall at Ardour University. In order to earn money necessary to remain at University she operates a fish and chip concession. Each evening she goes into town and buys packets of fish and chips which she brings back to the hall to resell. She pays 10p for the packets and sells them at 20p each. Any unsold packets are thrown away.
  Penny has kept a record of the demand (not actual sales) over the last 200 days, and her results were:

No. of packets demanded	No. of days when demanded	No. of packets demanded	No. of days when demanded
10	1	21	17
11	3	22	16
12	4	23	13
13	5	24	12
14	8	25	9
15	11	26	7
16	13	27	5
17	15	28	4
18	17	29	3
19	18	30	1
20	18		

By using relative frequencies to estimate probabilities:

(a) What is Penny's expected return if she always buys
   (i) 24 packets?
   (ii) 16 packets?

(b) By considering what Penny's returns would be if she always had exact prior knowledge of the demand, suggest and calculate a measure of the cost to her of, in fact, being uncertain of the demand.

9. Some general comments on expected value and variance. If a and b are two constants, and X is any random variable, the following important relationships always hold:

   (i) $E(aX + b) = aE(X) + b$.
   (ii) $\text{var} (aX + b) = a^2 \text{var} X$.

Taking the simple distribution of Section 2 by way of illustration, and a value of 3 for a and 2 for b:

Value of X (%)	$P(X = x)$	Value of $(aX + b)$ (%)
10	0·20	32
11	0·40	35
12	0·30	38
13	0·05	41
14	0·05	44

Thus,

$$E(3X + 2) = (32 \times 0·20) + (35 \times 0·40) + (38 \times 0·30)$$
$$+ (41 \times 0·05) + (44 \times 0·05)$$
$$= 36·05$$
$$= (3 \times 11·35) + 2$$
$$= 3E(X) + 2.$$

Using the same values for a and b:

Values of $(aX+b)(\%)$	Values of $(aX+b)-E(aX+b)$	Values of $\{(aX+b)-E(aX+b)\}^2$	Probability
32	−4·05	16·4025	0·20
35	−1·05	1·1025	0·40
38	1·95	3·8025	0·30
41	4·95	24·5025	0·05
44	7·95	63·2025	0·05

Now,

$$\begin{aligned}
\text{var}\,(3X+2) &= E\{(3X+2)-E(3X+2)\}^2 \\
&= (16\cdot4025 \times 0\cdot20) + (1\cdot1025 \times 0\cdot40) \\
&\quad + (3\cdot8025 \times 0\cdot30) + (24\cdot5025 \times 0\cdot05) \\
&\quad + (63\cdot2025 \times 0\cdot05) \\
&= 9\cdot2475 \\
&= 9 \times 1\cdot0275 \\
&= 9\,\text{var}\,X.
\end{aligned}$$

The addition of the same constant to all the values of a random variable does not affect the spread of the values. Consequently, since variance is a measure of dispersion, one could have anticipated that the right-hand side of (*ii*) would not depend on the value of b. Furthermore, since variance involves squaring the values of the random variable, it is not surprising that multiplication of those values by a, results in the variance increasing by a factor of a^2. The standard deviation, of course, is multiplied by just a.

SOME GENERAL COMMENTS ON DISTRIBUTIONS

The following remarks apply to both frequency distributions and the analogous probability distributions.

10. Symmetry. A distribution is said to be symmetrical if an axis can be drawn through the graph of the distribution, parallel to the vertical axis, in such a way that the two halves of the graph are mirror images of one another (Fig. 27).

In each example the broken line is the axis of symmetry.

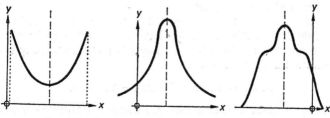

Fig. 27.—Some symmetrical probability distributions.

11. Modality. A "hump" in a distribution is called a mode, since it occurs at a most frequent value or range of values. The descriptions unimodal, bimodal, etc. refer in an obvious way to the number of modes (*see* Fig. 28).

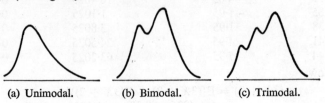

(a) Unimodal.   (b) Bimodal.   (c) Trimodal.

FIG. 28.—Modality of distributions.

12. Skewness. An asymmetrical distribution may be skewed to left or right (Fig. 29).

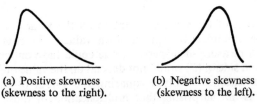

(a) Positive skewness   (b) Negative skewness
(skewness to the right).   (skewness to the left).

FIG. 29.—Skewness.

A general relationship between the mean, median, and mode holds for a skew distribution. Taking firstly a positively skewed distribution, the median is greater than the mode, since more than half the area beneath the curve lies to the right of the mode. Also, although half the area is to the right of the median it is spread over a greater length of the horizontal axis. As a result the mean is to the right of the median (Fig. 30).

An intuitive way of explaining this is to imagine the shape of the curve and horizontal axis to be cut out of a piece of wood. If an attempt is made to balance the shape on a knife edge at *A* it will topple down to the right. In order to obtain a balance, the knife edge must be placed at *B* (the mean).

A similar line of argument for a negatively skewed distribution shows that in ascending order one has mean, median, and mode. If the distribution is not too skewed an approximate relationship holds between the three statistics, *viz.*

$$\text{Mode} = \text{Mean} - 3 (\text{mean} - \text{median}).$$

Although measures of centrality and dispersion give a very useful summary of a distribution, skewness is a feature upon which they shed

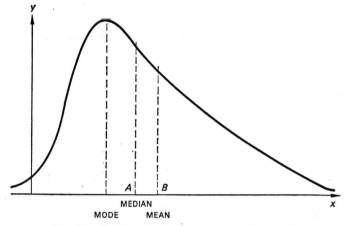

FIG. 30.—Relationship between measures of centrality.

no light. The two distributions of Fig. 31 both have the same mean and variance, for instance, but are differently skewed.

From the raw data it is possible, by finding the sign of the quantity (mean — median), to detect whether a distribution is negatively skewed (sign negative) or positively skewed (sign positive). However, it is not always easy to find an exact value for the median, and a second procedure may be based on calculating the *third moment* about the mean:

$$\int (x - \mu)^3 p(x)dx \text{ (continuous), or } \Sigma(x_i - \mu)^r p(x_i) \text{ (discrete).}$$

If the curve is symmetrical, this quantity is zero, since the positive contributions (arising from values of x and $p(x)$ to the right of the mean) are exactly matched by negative ones (from values of x the

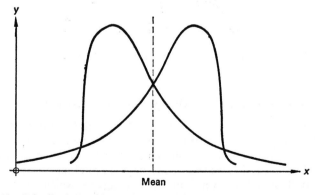

FIG. 31.—Distributions with equal means and variances, but oppositely skewed.

same distance to the left of the mean and with equal values of $p(x)$ because of symmetry). Negative skewness causes this third moment to be negative, and positive skewness causes it to be positive.

Exercises on Section 12

9. The *probability distribution function* $F(x)$ of any random variable X is defined by:

$$F(x) = \text{probability that } X \leqslant x.$$

The value of $F(x)$ is thus the area underneath the probability density function curve $f(x)$ to the left of a particular value, x (*see* Fig. 32).

(*a*) Explain why $0 \leqslant F(x) \leqslant 1$.
(*b*) Why does $F(x)$ never decrease as x increases?
(*c*) The probability distribution function of the Pareto distribution (*see* Exercises on Section 8, Question 6, above) is

$$F(x) = 1 - \left[\frac{x_0}{x}\right]^k$$

Use this fact to find the median income.

HINT: By definition, the area to the left of the median value is 0·50.

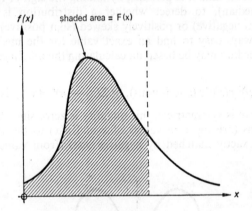

FIG. 32.—Probability distribution function.

MORE THAN ONE RANDOM VARIABLE

13. Joint distributions. The notion of joint probability was introduced in Chapter 3, and we can extend the ideas a little here in the light of the discussion of distributions. A random experiment in the most general sense can quite easily give rise to two or more random variables.

Selection of a tax-payer leads to two figures which may be of interest, his gross income and his net income. Similarly, selection of a company may be followed by consideration of share capital, loan stock, total assets, working capital, and profits after tax (*i.e.*, five random variables).

In order to fix our ideas, consider a third example of the two random variables (*i*) finished goods inventory (X) and (*ii*) work-in-process (Y). In studying their relationship it is useful to find the probability of one given the other, the expected value and variance of their sum, and so on. Accordingly, it is natural to think of a joint probability distribution which gives a breakdown of the various combinations of X and Y, together with their probabilities $P(X, Y)$. For simplicity suppose that the number of units of work-in-process can range between 0 and 2, and the number of units of inventory between 0 and 6. The joint probability table might be that shown in Table 20.

Table 20. Joint probability table of work-in-process and finished goods

Work-in-process	Finished Goods Inventory						
	0	1	2	3	4	5	6
0	0·00	0·02	0·03	0·04	0·04	0·06	0·07
1	0·03	0·05	0·05	0·06	0·06	0·07	0·08
2	0·07	0·06	0·05	0·05	0·04	0·04	0·03

Here, $P(X = 1, Y = 2)$, for example, is 0·06.

The marginal totals give the marginal distributions of X and Y (*i.e.*, the probabilities of various values of one variable occurring, irrespective of the value of the other):

$$P(X = 0) = 0·00 + 0·03 + 0·07 = 0·10$$
$$P(X = 1) = 0·13$$
$$P(X = 2) = 0·13$$
$$P(X = 3) = 0·15 \qquad \text{Marginal probability}$$
$$P(X = 4) = 0·14 \qquad \text{distribution of } X.$$
$$P(X = 5) = 0·17$$
$$P(X = 6) = 0·18$$
$$P(Y = 0) = 0·00 + 0·02 + \ldots + 0·07 = 0·26$$
$$P(Y = 1) = 0·40 \qquad \text{Marginal probability}$$
$$P(Y = 2) = 0·34 \qquad \text{distribution of } Y.$$

By direct extension of previously introduced ideas, one may talk of the conditional distribution of X given Y (and Y given X, for that matter), and define it by:

$$P(X|Y) = \frac{P(X, Y)}{P(Y)}.$$

Using this definition, the conditional distribution of X, given that $Y = 2$, is:

$$P(X = 0 \mid Y = 2) = \frac{P(X = 0 \text{ and } Y = 2)}{P(Y = 2)}$$

$$= \frac{0.07}{0.07 + \ldots + 0.03} = \frac{0.07}{0.34} = 0.2059$$

$$P(X = 1 \mid Y = 2) = 0.06/0.34 = 0.1765$$
$$P(X = 2 \mid Y = 2) = 0.05/0.34 = 0.1471$$
$$P(X = 3 \mid Y = 2) = 0.05/0.34 = 0.1471$$
$$P(X = 4 \mid Y = 2) = 0.04/0.34 = 0.1176$$
$$P(X = 5 \mid Y = 2) = 0.04/0.34 = 0.1176$$
$$P(X = 6 \mid Y = 2) = 0.03/0.34 = 0.0882$$

Furthermore, two random variables X, Y are defined as being *independent* if the conditional probability distributions of one, given each value of the second in turn, are all equal to the corresponding unconditional probability distributions, *i.e.* if

$$P(X = x \mid Y = y) = P(X = x)$$

for *all* values of X, Y. As a consequence of this definition, the independence of X and Y implies that

$$P(X = x, Y = y) = P(X = x) \times P(Y = y)$$

for all possible pairs of values (x, y).

So far the discussion has been restricted to the joint distribution of two discrete random variables. However, the ideas and definitions can be modified to cover continuous joint probability distributions. Such a distribution arises when X is, say, volume of liquid output of a chemical plant and Y is time. By analogy with the univariate case, a bivariate probability distribution can be visualised in three dimensions (Fig. 33).

The values of X and Y are given on the x, y axes, which are at right angles to each other in a plane perpendicular to the page, and the joint probability density function on the vertical axis. As a result of its dependence on two variables, the p.d.f. generates an undulating surface (as opposed to a curve) and probabilities are represented by volumes (rather than areas). The probability $P(0 \leqslant X \leqslant 2, 2 \leqslant Y \leqslant 3)$ would be equal to the volume underneath the surface between these limits (Fig. 33).

14. Sums of random variables. Using the example introduced in Section 13 above, the sum $(X + Y)$ is a quite legitimate random variable whose values are the total number of units already manufactured *and* in process of manufacture. As such it has an expected value which is easily calculated, bearing in mind that $(X + Y)$ can assume any value between 0 and 8 inclusive, and that most of the values

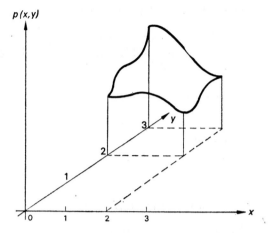

FIG. 33.—Graphical representation of joint probability density function.

can occur in several ways. For instance, $X + Y = 2$ can arise from $X = 1$, $Y = 1$ (with a probability of 0·05), $X = 0$, $Y = 2$ (with a probability of 0·07) or $X = 2$, $Y = 0$ (with a probability of 0·03). The value of $P(X + Y = 2)$ is therefore $0·05 + 0·07 + 0·03 = 0·15$.

The complete analysis is shown in Table 21.

Table 21. Probability table of random variable $(X + Y)$

Value of $(X + Y)$	Probability
0	0·00
1	0·02 + 0·03 = 0·05
2	0·05 + 0·07 + 0·03 = 0·15
3	0·06 + 0·05 + 0·04 = 0·15
4	0·05 + 0·06 + 0·04 = 0·15
5	0·05 + 0·06 + 0·06 = 0·17
6	0·04 + 0·07 + 0·07 = 0·18
7	0·04 + 0·08 = 0·12
8	0·03

Thus:

$$
\begin{aligned}
E(X + Y) &= (0 \times 0·00) + (1 \times 0·05) + (2 \times 0·15) + (3 \times 0·15) \\
&\quad + (4 \times 0·15) + (5 \times 0·17) + (6 \times 0·18) \\
&\quad + (7 \times 0·12) + (8 \times 0·03) \\
&= 0 + 0·05 + 0·30 + 0·45 + 0·60 + 0·85 + 1·08 \\
&\quad + 0·84 + 0·24 \\
&= 4·41.
\end{aligned}
$$

From the marginal distributions of X and Y we also have:

$$\begin{aligned}
E(X) &= (0 \times 0{\cdot}10) + (1 \times 0{\cdot}13) + (2 \times 0{\cdot}13) + (3 \times 0{\cdot}15) \\
&\qquad + (4 \times 0{\cdot}14) + (5 \times 0{\cdot}17) + (6 \times 0{\cdot}18) \\
&= 0{\cdot}13 + 0{\cdot}26 + 0{\cdot}45 + 0{\cdot}56 + 0{\cdot}85 + 1{\cdot}08 \\
&= 3{\cdot}33
\end{aligned}$$

and

$$\begin{aligned}
E(Y) &= (0 \times 0{\cdot}26) + (1 \times 0{\cdot}40) + (2 \times 0{\cdot}34) \\
&= 0{\cdot}40 + 0{\cdot}68 \\
&= 1{\cdot}08.
\end{aligned}$$

We can see immediately that

$$E(X + Y) = E(X) + E(Y). \tag{i}$$

The result that the expected value of the sum of two random variables is the sum of their individual expected values is a general one, and is not affected by the independence or dependence of the two variables. In fact the result can be generalised further, in that if $X_1, \ldots, X_n$ are any random variables,

$$E(X_1 + \ldots + X_n) = E(X_1) + \ldots + E(X_n). \tag{ii}$$

This property of expected values can be utilised to give the alternative expression var $X = E(X^2) - [E(X)]^2$ for the variance of a random variable.* This result may be interpreted verbally as the "expected value of the square (of the random variable) minus the square of the expected value (of the random variable)." The reader will note the strong analogy between this and the alternative formula for calculating the variance of a frequency distribution (*see* Chapter 1, Section 9).

---

* Multiplying out the right-hand side:

$$\text{var } X = E[X - E(X)]^2 = E\{X^2 - 2XE(X) + [E(X)]^2\}.$$

Since the expectation of this sum is the sum of the expectations, this becomes

$$\text{var } X = E(X^2) - E[2X \cdot E(X)] + E[E(X)]^2.$$

Now $E(X)$ is a constant, so

$$E[2XE(X)] = 2E(X) \cdot E(X) = 2[E(X)]^2$$

and

$$E[E(X)]^2 = [E(X)]^2.$$

Thus,

$$\begin{aligned}
\text{var } X &= E(X^2) - 2[E(X)]^2 + [E(X)]^2 \\
&= E(X^2) - [E(X)]^2 \tag{iii}
\end{aligned}$$

15. Covariance. Directing attention now to the random variable $(X + Y)$, we may use the final result of Section 14 above to obtain

$$\text{var}(X + Y) = \text{E}[(X + Y)^2] - [\text{E}(X + Y)]^2.$$

To calculate the first term on the right-hand side, in the context of the previous data, we need Table 22.

Table 22. Probability table of random variable $(X + Y)^2$

Value of $(X + Y)^2$	Probability
0	0·00
1	0·05
4	0·15
9	0·15
16	0·15
25	0·17
36	0·18
49	0·12
64	0·03

By definition of expectation,

$$\begin{aligned}
\text{E}\{(X + Y)^2\} = {} & (0 \times 0\cdot00) + (1 \times 0\cdot05) + (4 \times 0\cdot15) \\
& + (9 \times 0\cdot15) + (16 \times 0\cdot15) + (25 \times 0\cdot17) \\
& + (36 \times 0\cdot18) + (49 \times 0\cdot12) + (64 \times 0\cdot03) \\
= {} & 22\cdot93.
\end{aligned}$$

Now,

$$\text{E}(X + Y) = 4\cdot41 \text{ is already known, so}$$
$$\begin{aligned}
\text{var}(X + Y) &= 22\cdot93 - (4\cdot41)^2 \\
&= 22\cdot93 - 19\cdot45 \\
&= 3\cdot48.
\end{aligned}$$

Now,

$$\begin{aligned}
\text{var } X = {} & (0 \times 0\cdot10) + (1 \times 0\cdot13) + (4 \times 0\cdot13) \\
& + (9 \times 0\cdot15) + (16 \times 0\cdot14) + (25 \times 0\cdot17) \\
& + (36 \times 0\cdot18) - (3\cdot33)^2 \\
= {} & 14\cdot97 - 11\cdot09 \\
= {} & 3\cdot88
\end{aligned}$$

and

$$\begin{aligned}
\text{var } Y &= (0 \times 0\cdot26) + (1 \times 0\cdot40) + (4 \times 0\cdot34) - (1\cdot08)^2 \\
&= 0\cdot59.
\end{aligned}$$

Thus

$$\text{var}(X + Y) \neq \text{var } X + \text{var } Y.$$

The relationship between the variance of a sum of random variables, and the variances of the components is not the same as with expected values. In fact, var $(X + Y) =$ var $X +$ var Y only if X and Y are independent random variables. In the illustration X and Y are not independent, therefore var $(X + Y) \neq$ var $X +$ var Y. Certainly there is a relationship between the variances, but the complete picture also involves a measure of that part of the total variation which is due to large (small) values of X tending to occur with large (small) values of Y. This measure is called the *covariance of X and Y* (written cov (X, Y)), and is such that

$$\text{var}\,(X + Y) = \text{var}\,X + \text{var}\,Y + 2\,\text{cov}\,(X, Y). \qquad (iv)$$

This can be interpreted as defining cov (X, Y), since all the other quantities in the equation can be easily calculated. Notice that when X and Y are independent, cov $(X, Y) = 0$.

In general, if $X_1, \ldots, X_n$ are a mutually independent set of random variables:

$$\text{var}\,(X_1 \pm \ldots \pm X_n) = \text{var}\,X_1 + \ldots + \text{var}\,X_n.$$

Notice that the terms on the right-hand side of the equation are all positive, irrespective of whether the random variables are added together or subtracted.

Thus,

$$\text{var}\,(X_1 - X_2) = \text{var}\,(X_1 + X_2) = \text{var}\,X_1 + \text{var}\,X_2$$
$$\text{var}\,(X_1 - X_2 + X_3) = \text{var}\,(X_1 - X_2 - X_3)$$
$$= \text{var}\,X_1 + \text{var}\,X_2 + \text{var}\,X_3, \text{ etc.}$$

The same comment applies to the variance terms when the random variables are not independent, but in that case the sign of the co-variance term is affected.

Thus,

$$\text{var}\,(X_1 + X_2) = \text{var}\,X_1 + \text{var}\,X_2 + 2\,\text{cov}\,(X_1, X_2)$$

but

$$\text{var}\,(X_1 - X_2) = \text{var}\,X_1 + \text{var}\,X_2 - 2\,\text{cov}\,(X_1, X_2).$$

Worked Example

By squaring out the terms on the right-hand side of the equation:

$$\text{var}\,(X + Y) = \text{E}(X + Y)^2 - [\text{E}(X + Y)]^2$$

and by using equation (iv) of Section 15 above, show that

$$\text{cov}\,(X, Y) = \text{E}\{[X - \text{E}(X)]\,[Y - \text{E}(Y)]\} = \text{E}(XY) - \text{E}(X)\,.\,\text{E}(Y).$$

Answer

$$\text{var}\,(X+Y) = E(X^2 + 2XY + Y^2) - [E(X) + E(Y)]^2$$
$$= E(X^2) + 2E(XY) + E(Y^2) - \{[E(X)]^2 + 2E(X)E(Y)$$
$$+ [E(Y)]^2\}$$
$$= E(X^2) - [E(X)]^2 + E(Y^2) - [E(Y)]^2$$
$$+ 2[E(XY) - E(X)E(Y)]$$
$$= \text{var}\,X + \text{var}\,Y + 2[E(XY) - E(X).E(Y)]. \qquad (i)$$

Now,

$$\text{var}\,(X+Y) = \text{var}\,X + \text{var}\,Y + 2\,\text{cov}\,(X,\,Y). \qquad (ii)$$

Thus, comparing (*i*) and (*ii*):

$$\text{cov}\,(X,\,Y) = E(XY) - E(X)E(Y).$$

Using the data provided:

XY		P(XY)	XYP(XY)
0	(from $X = 0$ or $Y = 0$)	$0 \cdot 10 + 0 \cdot 26 = 0 \cdot 36$	0·00
1	(from $X = 1$, $Y = 1$)	0·05	0·05
2	(from $X = 1$, $Y = 2$ and $X = 2$, $Y = 1$)	$0 \cdot 06 + 0 \cdot 05 = 0 \cdot 11$	0·22
3	(from $X = 3$, $Y = 1$)	0·06	0·18
4	(from $X = 4$, $Y = 1$ and $X = 2$, $Y = 2$)	$0 \cdot 06 + 0 \cdot 05 = 0 \cdot 11$	0·44
5	(from $X = 5$, $Y = 1$)	0·07	0·35
6	(from $X = 6$, $Y = 1$ and $X = 3$, $Y = 2$)	$0 \cdot 08 + 0 \cdot 05 = 0 \cdot 13$	0·78
7	(impossible)	0·00	0·00
8	(from $X = 4$, $Y = 2$)	0·04	0·32
9	(impossible)	0·00	0·00
10	(from $X = 5$, $Y = 2$)	0·04	0·40
11	(impossible)	0·00	0·00
12	(from $X = 6$, $Y = 2$)	0·03	0·36
Total		1·00	3·10

Thus, $E(XY) = 3 \cdot 10$. Also, $E(X) = 3 \cdot 33$ (*see* Section 14 above), $E(Y) = 1 \cdot 08$. Finally, $\text{cov}\,(X,\,Y) = 3 \cdot 10 - (3 \cdot 33)\,(1 \cdot 08) = -0 \cdot 50$.

Exercises on Section 15

10. Table 23 gives information on the gross income and tax liabilities of ten individuals.

Although the results discussed in the latter part of this chapter have been couched in terms of random variables they apply equally well to empirical data. Since each value of X occurs only once the relative frequency of each is 1/10 and in the absence of any other information we may take this figure as being the probability of occurrence of each of the values.

(*a*) Show that mean $(X - Y) = $ mean $X - $ mean Y.

(*b*) Calculate var X in two different ways, by analogy with the equation for random variables: var $X = E\{X - E(X)\}^2 = E\{E(X^2)\} - \{E(X)\}^2$ and verify that the results are equal.

Table 23. Tax information on 10 individuals

(1) Individual	(2) Gross Income (£) (X)	(3) Tax payable (£) (X − Y)	(4) Net Income (£) (Y)	(5) Tax payable ($) £1 = $2·55 2·55(X − Y)	(6) Net Income (allowances £800, tax rate 30%) (0·7 X −560)
A	3800	800	3000	2040·00	2100·00
B	2600	450	2150	1147·50	1260·00
C	4200	1000	3200	2550·00	2380·00
D	2950	700	2250	1785·00	1505·00
E	1895	295	1600	752·25	766·50
F	2125	325	1800	828·75	927·50
G	1976	293	1683	747·15	823·20
H	2005	118	1887	300·90	843·50
I	4018	518	3500	1320·90	2252·60
J	3236	389	2847	991·95	1705·20

(c) What would you anticipate to be the relationship between (i) the mean values and (ii) the variances of the figures in columns (3) and (5)? Check your assertion by calculation.

(d) What would you anticipate to be the relationship between (i) the mean values and (ii) the variances of the figures in columns (2) and (6)? Check your assertion by direct calculation.

(e) Calculate cov (X, Y) in three ways by analogy with the equation for random variables:

$$\text{cov}(X, Y) = E\{[X − E(X)][Y − E(Y)]\}$$
$$= E(XY) − E(X)E(Y)$$
$$= \tfrac{1}{2}\text{var}(X + Y) − \text{var } X − \text{var } Y.$$

Verify that the results are all equal.

(f) Suggest a means of calculating the third moment about the mean for grouped data. Calculate a value for this moment using the data provided, and comment on the significance of the result with regard to skewness.

CHAPTER 5

DISCRETE PROBABILITY MODELS

INTRODUCTION

In this chapter we shall examine several probability distributions which are of particular practical importance. The first few discrete distributions are introduced in the context of probabilistic depreciation, although other applications of the distributions are illustrated in the exercises at the end of the various sections.

SOME SPECIAL DISCRETE PROBABILITY MODELS

1. Bernoulli distribution. This is perhaps the simplest type of discrete distribution, and is associated with a random variable that can assume one of only two possible values with probabilities p, $(1 - p)$ respectively. Suppose the probability that a particular asset is retired in the first year of service is p. The random variable X, "number of assets retired after one year," can assume one of the two values, 0 or 1, with probabilities respectively of $(1 - p)$ and p. Notice that the sample space contains only two elementary events, the sum of whose probabilities must therefore be 1. This type of random variable is said to have a *Bernoulli* distribution. Its expected value and variance can be readily calculated:

Value of X	Probability
0	$(1 - p)$
1	p

Therefore,

$$E(X) = \{0 \times (1 - p)\} + (1 \times p)$$
$$= p.$$
$$\text{var } X = E(X^2) - \{E(X)\}^2$$
$$= [\{0^2 \times (1 - p)\} + (1^2 \times p)] - p^2$$
$$= p - p^2$$
$$= p(1 - p).$$

2. Binomial distribution. More generally, suppose we now have n identical assets, and that their service lives are all independent of each other. Of immediate interest is the probability, $P(r)$, that r of the assets are retired during the first year, if the probability of any individual asset being retired is p. The probabilities of the two extreme values of r—$P(0)$ and $P(n)$—are not too difficult to find, since the independence of the assets' behaviour enables the multiplicative rule (*see* Chapter 3, Section 2) to be used:

$$P(n) = \underbrace{p \times \ldots \times p}_{n \text{ times}} = p^n$$

$$P(0) = \underbrace{(1-p) \times \ldots \times (1-p)}_{n \text{ times}} = (1-p)^n.$$

For a general value of r between 0 and n the derivation of $P(r)$ requires a little more thought. If the n assets are labelled $a_1, \ldots, a_n$ and retirement and non-retirement at the end of one year are denoted by R, N respectively, the probability of a particular sequence involving r retirements, say

$$
\begin{array}{ccccc}
a_1 & a_2 & a_3 & a_4 \ldots a_n \\
R & R & N & R \ldots N
\end{array}
$$

can again be found, using the multiplicative rule, as:

$$p \times p \times (1-p) \times p \times \ldots \times (1-p).$$

This has the value $p^r(1-p)^{n-r}$, since there are r factors of p and $(n-r)$ factors of $(1-p)$ in the product. However, since the assets are all identical the point of interest is not in the probability of a particular sequence of retirements and non-retirements, but rather in just the occurrence of r retirements, irrespective of order.

Consequently, we need to know the number of ways exactly r retirements can occur amongst the n assets. Mathematically this is no different from selecting r machines for retirement out of the n, and we know from Chapter 3, Section 8, that the number of these combinations is

$$_nC_r = \frac{n!}{(n-r)!\,r!}.$$

Each of these combinations can occur with a probability of $p^r(1-p)^{n-r}$, since application of the multiplicative rule involves r p's and $(n-r)$ $(1-p)$'s in each case. Furthermore, the sequences are mutually exclusive since obviously the occurrence of one precludes the occurrence of any other. In terms of sample spaces then, the event "r machines are retired during first year" is made up of $_nC_r$ elementary events, each occurring with a probability of $p^r(1-p)^{n-r}$.

Thus,

$$P(r) = \underbrace{p^r(1-p)^{n-r} + \ldots + p^r(1-p)^{n-r}}_{{}_nC_r \text{ times}}$$

$$= \frac{n!}{r!(n-r)!} p^r(1-p)^{n-r}.$$

This result gives the complete probability distribution of the number of retirements at the end of the year, and the probability of any particular number, r, is found by inserting the appropriate values of n, p, and r in the formula. Notice in particular that

$$P(0) = \frac{n!}{0!\,n!} p^0(1-p)^n = (1-p)^n$$

and

$$P(n) = \frac{n!}{n!\,0!} p^n(1-p)^0 = p^n$$

which agree with the previous results.

For a more detailed illustration, suppose that $n = 4$, and $p = 0.2$. Then

$$P(0) = \frac{4!}{0!\,4!}\,(0.2)^0\,(0.8)^4 = 1(0.4096) \quad\;\;\, = 0.4096$$

$$P(1) = \frac{4!}{1!\,3!}\,(0.2)^1\,(0.8)^3 = 4(0.2)(0.5120) = 0.4096$$

$$P(2) = \frac{4!}{2!\,2!}\,(0.2)^2\,(0.8)^2 = 6(0.04)(0.64) \;\; = 0.1536$$

$$P(3) = \frac{4!}{3!\,1!}\,(0.2)^3\,(0.8)^1 = 4(0.008)(0.8) \;\; = 0.0256$$

$$P(4) = \frac{4!}{4!\,0!}\,(0.2)^4\,(0.8)^0 = 1(0.0016) \quad\;\;\, = 0.0016$$

By way of a check, $P(0) + P(1) + P(2) + P(3) + P(4) = 1$, as it should, since these events form an exhaustive and mutually exclusive set (*i.e.*, there must be either 0, 1, 2, 3, or 4 retirements during the first year, and only one of these results will occur).

3. Some general comments on the binomial distribution. The essential features of the binomial distribution can be stated in a general manner which enables the distribution to be used in a wide variety of circumstances. It is appropriate in any situation which involves n independent trials of a random experiment, with two possible outcomes at each trial: "success" (with probability p) and "failure" (with probability $(1 - p)$. It should be emphasised that the trials must all be independent, and that the probability of success (and therefore of failure) does not vary between any of the trials. "Success" and "failure" are to be liberally interpreted as referring to the two possible outcomes, and should not be taken in their literal sense.

In our previous example the n trials were the performances of the n machines during the year, and the two possible outcomes were retirement (success) and non-retirement (failure).

The simplest way to obtain the expected value and variance of the binomial random variable is to consider the random variable R, the number of successes out of n trials, as the sum of n independent Bernoulli random variables. Any particular trial can have one of two possible outcomes, success or failure, with probabilities p, $(1 - p)$ respectively. The total number of "successes" (or retirements) amongst n trials (or assets) is therefore determined by:

$$R = X_1 + \ldots + X_n$$

where X_i is a Bernoulli random variable having value one if the ith trial is a success, and value zero if the ith trial results in failure.

Also,

$$E(R) = E(X_1) + \ldots + E(X_n).$$

Each Bernoulli random variable has an expected value of p, therefore

$$E(R) = np.$$

Furthermore, since $X_1, \ldots, X_n$ are independent,

$$\begin{aligned} \text{var } R &= \text{var } (X_1 + \ldots + X_n) \quad (\textit{see Chapter 4, Section 15}). \\ &= \text{var } X_1 + \ldots + \text{var } X_n \\ &= np(1 - p). \end{aligned}$$

The value for $E(R)$ has a simple commonsense interpretation. If we repeated a task 50 times, $\textit{i.e. } n = 50$, and the probability of success each time was 0.3, we would expect to be successful roughly

$$0.3 \times 50 = 15 \text{ times.}$$

The probability of exactly r successes may not always be the value of central interest. It may be useful, for example, to know the probability that there are at least 2 successes, or less than 3 successes. There is no general short cut to finding these values exactly—one must turn to the detailed probability distribution.

With $n = 6$, and $p = 0.3$ as before:

$$P(R \geqslant 2) = P(R = 2) + P(R = 3) + P(R = 4) \\ + P(R = 5) + P(R = 6) \\ = 0.3241 + 0.1852 + 0.0595 + 0.0102 + 0.0007 \\ = 0.5797$$

and

$$P(R < 3) = P(0) + P(1) + P(2) \\ = 0.1177 + 0.3025 + 0.3241 \\ = 0.7443.$$

A useful device for finding the first of these values lies in recognising that $\{R < 2\}$ and $\{R \geqslant 2\}$ are complementary events.

This being so,

$$P(R < 2) + P(R \geqslant 2) = 1$$

and

$$P(R \geqslant 2) = 1 - P(R < 2).$$

Furthermore, $P(R < 2)$ is just $P(R = 0) + P(R = 1)$.

Therefore,

$$P(R \geqslant 2) = 1 - \{P(R = 0) + P(R = 1)\} \\ = 1 - (0.1177 + 0.3025) \\ = 0.5798$$

as before. This approach can often be used to reduce the amount of computation necessary in finding probabilities.

The binomial distribution is illustrated in Fig. 34 for different values of n and p. Unimodality is a general feature of the distribution, although the value of the mode depends upon the relative values of n and p. If p is near 0.5 the distribution is roughly symmetrical, whilst for values of p near 0 or 1 it is noticeably skewed.

For any probability of success, p, the distribution is the mirror image of that for a probability of success of $(1 - p)$.

Worked Examples

1. A statistics test consists in providing Yes/No answers to ten questions. To pass the test a candidate must provide at least 4 correct answers. Unbeknown to the invigilator a well-behaved chimpanzee sneaks into the examination and answers the questions randomly

$$(i.e., P(\text{right}) = P(\text{wrong}) = \tfrac{1}{2}).$$

What is the probability that the chimpanzee scores a passable mark?

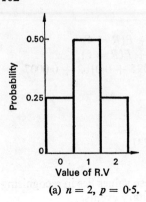

(a) $n = 2$, $p = 0.5$.

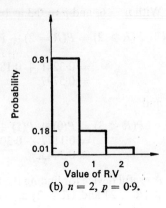

(b) $n = 2$, $p = 0.9$.

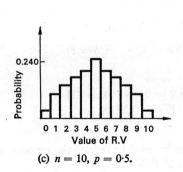

(c) $n = 10$, $p = 0.5$.

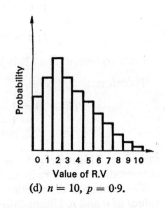

(d) $n = 10$, $p = 0.9$.

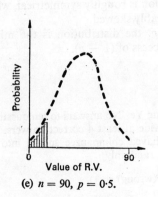

(e) $n = 90$, $p = 0.5$.

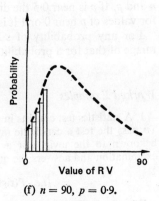

(f) $n = 90$, $p = 0.9$.

Fig. 34.—Examples of the binomial distribution.

Answer

Let R be the number of correct answers obtained. We require

$$P(R \geqslant 4) = 1 - P(R < 4)$$

Now

$$P(R < 4) = P(0) + P(1) + P(2) + P(3)$$

Also,

$$P(0) = \frac{10!}{0!\,10!}\,(\tfrac{1}{2})^0\,(\tfrac{1}{2})^{10} = (\tfrac{1}{2})^{10}$$

$$P(1) = \frac{10!}{1!\,9!}\,(\tfrac{1}{2})^1\,(\tfrac{1}{2})^9 = 10(\tfrac{1}{2})^{10}$$

$$P(2) = \frac{10!}{2!\,8!}\,(\tfrac{1}{2})^2\,(\tfrac{1}{2})^8 = 45(\tfrac{1}{2})^{10}$$

$$P(3) = \frac{10!}{3!\,7!}\,(\tfrac{1}{2})^3\,(\tfrac{1}{2})^7 = 120(\tfrac{1}{2})^{10}$$

Thus,

$$P(R < 4) = (1 + 10 + 45 + 120)(\tfrac{1}{2})^{10}$$
$$= 176/1024.$$

Finally,

$$P(R \geqslant 4) = 1 - \frac{176}{1024} = \frac{848}{1024} = 0\cdot8281.$$

The chimpanzee has a perhaps surprisingly large probability of passing the test! Notice that the calculations were quite simple in this particular case because the probabilities of "success" and "failure" were equal.

2. An entrepreneur simultaneously opens an exclusive ladies' boutique in each of eight cities. Assuming that each boutique has a probability of $0\cdot3$ of realising a profit in the first year, and that these probabilities are all independent, what is the probability that

    (*a*) at least three realise a profit;
    (*b*) exactly three realise a profit;
    (*c*) more than six show a loss.

Answer

Let B be the number of boutiques showing a profit.

    (*a*) We require

$$P(B \geqslant 3) = 1 - P(B < 3)$$
$$= 1 - \{P(B = 0) + P(B = 1) + P(B = 2)\}.$$

Now

$$P(B = 0) = \frac{8!}{0!\,8!}\,(0\cdot3)^0\,(0\cdot7)^8 = 0\cdot0576$$

$$P(B = 1) = \frac{8!}{1!\,7!}\,(0\cdot3)^1\,(0\cdot7)^7 = 0\cdot1977$$

$$P(B = 2) = \frac{8!}{2!\,6!}\,(0\cdot3)^2\,(0\cdot7)^6 = 0\cdot2965.$$

So,

$$P(B \geqslant 3) = 1 - 0.5518$$
$$= 0.4482.$$

(b) $P(B = 3) = \dfrac{8!}{3!\,5!}\,(0.3)^3\,(0.7)^5 = 0.2541.$

(c) $P(B < 2) = P(B = 0) + P(B = 1)$
$$= 0.2553.$$

Exercises on Section 3

1. An insurance company knows from previous records that the probability of a person in a particular age-group dying in the next year is 0·15. It has 9 policy-holders who fall into this age range. What is the probability that the company will receive

(a) more than one;
(b) less than seven;
(c) exactly four;

claims in the next year, as a result of deaths within this group?

2. A stockbroker claims to be able to separate new issues into high and low growth stocks, according to some predetermined criteria. Given six new issues he correctly classifies four, and misclassifies two. What is the probability of obtaining a score as good as this purely by chance? In the light of your answer, what credibility do you give to the broker's claim?

3. A market research organisation carries out a pilot test involving the mailing of questionnaires to 7 individuals. From previous experience it is estimated that the probability of an individual returning a completed questionnaire is 0·25. Find the probability that at most one individual makes a return.

4. Ten stocks are selected by an investor for their excellent past performance. Their prices are assumed to move independently, and the probability of any one of them showing a fall in price at the end of the coming month is estimated at 0·04. What is the probability that more than one stock shows a price fall over this period?

5. An accountant is to audit 24 accounts of a firm. Sixteen of these are of highly-valued customers. If the accountant selects four of the accounts at random, what is the probability that he chooses at least one highly-valued account?

4. Geometric distribution. A random variable, S, which can assume any of the values 1, 2, 3, ..., with probabilities $P(s)$, where

$$P(s) = P(S = s) = (1 - p)^{s-1}p \qquad (0 < p < 1) \quad (i)$$

is said to have a *geometric* distribution.

We note that p is simply interpreted as the probability that $S = 1$ since,

$$P(S = 1) = (1 - p)^{1-1}p$$
$$= p.$$

Furthermore, we can see that the probabilities decrease by the same factor of $(1 - p)$ for each unit increase in the value of S. The random variable S is an example of the discrete type, but which can theoretically assume any one of an infinite number of possible values.

That (i) does in fact represent a legitimate probability distribution can be easily checked by showing that the sum of the probabilities of all the elementary events is unity. Now,

$$\text{Total probability} = P(1) + P(2) + \ldots$$
$$= p + (1 - p)p + (1 - p)^2 p + \ldots$$
$$= p\{1 + (1 - p) + (1 - p)^2 + \ldots\}.$$

Since $(1 - p)$ is less than one, the sum in brackets { } can be recognised as the sum of an infinite number of terms of a geometric series with first term 1 and common ratio $(1 - p)$.

Accordingly,

$$\text{Total probability} = p\left\{\frac{1}{1 - (1 - p)}\right\}$$
$$= p/p = 1$$

as required.

The calculation of the expected value, $E(S)$ and variance, var (S), of the geometric distribution requires more advanced knowledge of series and differentiation. We simply quote the results here as $\frac{1}{p}$, and $\frac{1}{p}\left(\frac{1}{p} - 1\right)$ respectively. Some examples of the shape of the geometric distribution for differing values of p are shown in Fig. 35.

The service-life, S years, of an asset is a random variable, since on acquisition it is not usually known how long it will survive. To express this uncertainty a geometric distribution can be used, the value of the constant p being chosen within the limits $0 < p < 1$ to suit the particular character of the asset.

Certainly the asset will be retired at some time, and it could be argued that the use of such a model is not realistic since S can take any integer value. However, since p, and therefore $(1 - p)$, is less than one, $P(s)$ decreases as s increases, and rapidly becomes negligibly small. This is certainly what one would look for in any model of an asset's service life, and the value of the geometric model here rests on its theoretical simplicity and its very good approximation to practical situations.

By way of illustration, suppose that the probability of the asset's retirement at the end of one year is 0·3. The distribution of service-lives can be modelled by:

$$P(1) = 0·3$$
$$P(2) = 0·3 \times 0·7 = 0·21$$
$$P(3) = 0·147$$
$$P(4) = 0·1029, \text{ etc.}$$

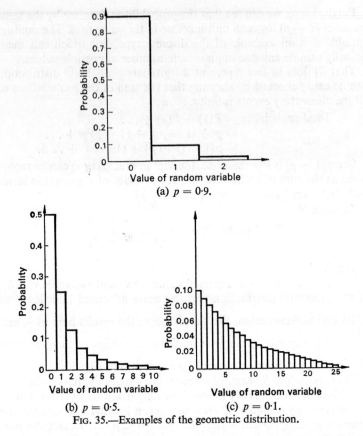

(a) $p = 0.9$.

(b) $p = 0.5$.

(c) $p = 0.1$.

FIG. 35.—Examples of the geometric distribution.

and the expected value and standard deviation of service-lives are $1/0.3 = 3.33$ years, and

$$\sqrt{\left\{ \frac{1}{0.3} \left(\frac{1}{0.3} - 1 \right) \right\}} = 2.79 \text{ years respectively.}$$

Worked Example

Suppose that the probability of an asset of a particular type being retired after one year is 0.1. If five assets of this type are acquired give, at the time of acquisition, the complete probability distribution of the number of retirements that have taken place by the end of two years.

Answer

Since $p = 0.1$, then for a particular asset:

$$P(\text{retirement by end of two years}) = P(1) + P(2)$$
$$= 0.10 + 0.09$$
$$= 0.19.$$

Now, the events "retirement by end of two years" and "non-retirement by the end of two years" represent two exhaustive and exclusive ways in which each of the five assets can behave. The probabilities of 0, 1, 2, 3, 4, 5 retirements by the end of two years are therefore given by the binomial distribution with $n = 5$, $p = 0.19$ (and $1 - p = 0.81$):

$$P(0) = \frac{5!}{0!\,5!}\,(0.19)^0\,(0.81)^5 = 0.3487$$

$$P(1) = \frac{5!}{1!\,4!}\,(0.19)^1\,(0.81)^4 = 0.4089, \text{ etc.}$$

Exercises on Section 4

6. The probability that an asset will be retired after one year is 0.2. Use the geometric distribution to find the probability that it will have a service life of (*i*) at least five years; (*ii*) just four years; (*iii*) no more than three years.

7. A mining company is about to begin explorations for copper in a new area. The probability of finding ore with an economic copper content in the first two months of operations is estimated, on the basis of geological information, to be 0.7. Use a geometric model to find the probability that no suitable ore has been found after 6 months.

5. The Poisson distribution. There are many instances of particular events occurring at random—insurance claims, breakdowns of machines, errors in a ledger, and so on. As in the case of all these examples, the randomness is usually in space or time. The number of events in a particular interval of space or time (such as the number of flaws per 10 metres of cloth, or the number of arrivals in a queue per minute) is a random variable which may assume any of the values 0, 1, 2, ...

With the assumptions that:

(*a*) the occurrence or non-occurrence of an event does not influence the occurrence or non-occurrence of any other,

(*b*) the probability of one event occurring in a very small interval of space or time δt is proportional to the length of that interval (in other words this probability is $\lambda \delta t$ where λ is a constant),

(*c*) the probability of more than one event happening in a very small interval is negligible,

the probability of r events occurring in interval t is

$$P(r) = \frac{e^{-\lambda t}\,(\lambda t)^r}{r!} \qquad r = 0, 1, 2, \ldots.$$

This distribution is called a *Poisson distribution*, and the circumstances which give rise to it a *Poisson process*. The values of $P(r)$ are dependent on λ and t, and these two quantities must be prescribed in order to define the distribution completely.

Clearly there exists a family of Poisson distributions corresponding

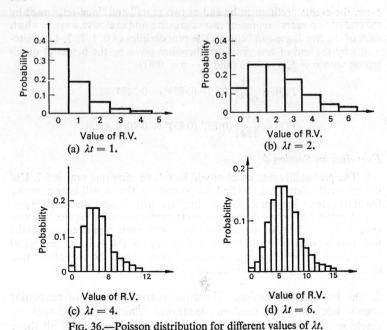

Fig. 36.—Poisson distribution for different values of λt.

to different values of λ and t. It is possible to show that for any random variable R, with a Poisson distribution defined by λ, t that:*

(i) Total probability $= P(0) + P(1) + \ldots$

$$= e^{-\lambda t} + \frac{e^{-\lambda t}\,\lambda t}{1!} + \frac{e^{-\lambda t}\,(\lambda t)^2}{2!} + \ldots = 1$$

(ii) $$\mathrm{E}(R) = 0 \cdot e^{-\lambda t} + \frac{1 \cdot e^{-\lambda t}\,\lambda t}{1!} + \frac{2 \cdot e^{-\lambda t}\,(\lambda t)^2}{2!} + \ldots$$
$$= \lambda t$$

(iii) $$\mathrm{var}\,R = \mathrm{E}(R^2) - [\mathrm{E}(R)]^2 = \lambda t.$$

The expected value and variance of a Poisson random variable are therefore equal. The fact that $\mathrm{E}(R) = \lambda t$ enables us conveniently to interpret λ as the average rate of occurrence of events, since λ is the expected number in interval t.

The Poisson distribution is illustrated in Fig. 36 for several different values of the parameter, λt, on which it depends.

* The reader may recognise the sum $\sum_{r=0}^{\infty} e^{-\lambda t}\,(\lambda t)^r / r! = e^{-\lambda t} \sum_{r=0}^{\infty} (\lambda t)^r / r!$ on the right-hand side of (i) as the expansion of $e^{\lambda t}$ in ascending powers of λt. Thus total probability $= e^{-\lambda t} \cdot e^{\lambda t} = e^0 = 1$. (ii) and (iii) are derived similarly.

Consider, for example, the estimation of the costs of breakdowns in a machine shop. As part of this exercise it is essential to have available a model of the pattern of breakdowns. Since breakdowns occur independently and at random (to a good approximation) the Poisson process is an appropriate theoretical model to use. Suppose that the number of breakdowns has averaged 2 per week in the past, and the distribution of breakdowns over the next three-week period is required. It is imperative that λ and t are measured in compatible units—here the average rate of occurrence of breakdown is given per week, so $\lambda = 2$, and the interval of interest is three weeks, so $t = 3$. There is no reason why λ should not be defined as 2/5 per day (assuming a five-day week), but then $t = 15$ days. In both cases the product λt, which appears in the equation for $P(r)$, has the value 6. The complete distribution of breakdowns over the period can now be specified:

$$P(r) = e^{-6} (6)^r/r! \qquad r = 0, 1, 2, \ldots$$

e.g.,
$$P(0) = e^{-6} (6)^0/0!$$
$$= e^{-6}$$
$$= 0 \cdot 00248$$
$$P(1) = e^{-6} (6)^1/1!$$
$$= 6e^{-6}$$
$$= 0 \cdot 01488$$
$$P(2) = e^{-6} (6)^2/2!$$
$$= 18e^{-6}$$
$$= 0 \cdot 04464, \text{ etc.}$$

Worked Examples

1. The purchasing manager of a department store knows from past experience that the demand for a particular article of furniture averages two per day. The wholesaler is able to deliver the articles to the store (before it opens) on Wednesday and Friday mornings only. On Tuesday evening, the purchasing manager—on discovering that he is out of stock—orders four articles for delivery the following morning. He reasons that the probability of his having sufficient stocks until Thursday evening is 0·5.

(a) On the assumption that demand is random, give full calculations to show why you agree or disagree with his reasoning.
(b) If the only cost of holding stocks of the article in the store is 75p per article per night, what is the expected cost of storage of the articles for Wednesday night?

Answer

(a) Average demand per day, $\lambda = 2$.
Number of days (Weds., Thurs.), $t = 2$.

Probability that stocks will last until Thursday evening =
Probability that demands number less than or equal to 4

$$= e^{-2.2} + \frac{e^{-2.2}(2.2)^1}{1!} + \frac{e^{-2.2}(2.2)^2}{2!} + \frac{e^{-2.2}(2.2)^3}{3!} + \frac{e^{-2.2}(2.2)^4}{4!}$$

$$= e^{-4} + 4e^{-4} + 8e^{-4} + \frac{64}{6}e^{-4} + \frac{256}{24}e^{-4}$$

$$= \frac{103}{3}e^{-4}$$

$$= 0.6283.$$

He has underestimated the probability that he will have sufficient stocks.

(b) Let $P(r)$ be the probability that r articles are stored on Wednesday night. Then:

Expected number stored =

$$0.P(0) + 1.P(1) + 2.P(2) + 3.P(3) + 4.P(4)$$

Furthermore, there will be 4 to store on Wednesday night if there are no demands during Wednesday. Thus

$$P(4) = e^{-2}$$

(note that $t = 1$, $\lambda = 2$ now, since we are concerned with one day—Wednesday—only).

Similarly,

$$P(3) = 2e^{-2}, \; P(2) = 2e^{-2}, \; P(1) = \tfrac{4}{3}e^{-2}.$$

However, $P(0)$ will be equal to the probability that there are 4 *or more* demands on Wednesday.

Accordingly,

$$P(0) = 1 - \left\{ e^{-2} + \frac{e^{-2} \cdot 2}{1!} + \frac{e^{-2} \cdot 2^2}{2!} + \frac{e^{-2} \cdot 2^3}{3!} \right\}$$

$$= 1 - e^{-2}\{2 + 2 + \tfrac{4}{3}\}$$

$$= 1 - \frac{16}{3}e^{-2}$$

$$= 0.2784.$$

Finally,

expected number stored $= (0 \times 0.2784) + (1 \times 0.1804) + (2 \times 0.2706)$
$$+ (3 \times 0.2706) + (4 \times 0.1353)$$
$$= 2.075$$

and,

expected storage costs $= 2.075 \times 75p$
$$= £1.56.$$

2. An executive makes on average five telephone calls per hour, at a cost which may be taken as 5p per call. Determine the probability that in any hour the telephone costs exceed 15p.

Answer
Here $\lambda = 5$ per hour, and $t = 1$ hour.

Thus,

Probability that
costs exceed 15p = Probability of more than three calls

$$= 1 - \left\{ e^{-5} + \frac{e^{-5} \cdot 5^1}{1!} + \frac{e^{-5} \cdot 5^2}{2!} + \frac{e^{-5} \cdot 5^3}{3!} \right\}$$

$$= 1 - \frac{236}{6} e^{-5}$$

$$= 0 \cdot 7350.$$

Exercises on Section 5

8. If the number of breakdowns in a large machine shop averages three per week, what is the probability of more than six breakdowns occurring in a two-week period?

9. The number of entries made in each of five accounts receivable averages 1·5 per day, and entries may be assumed to occur independently. What is the probability of a particular account having:

    (*a*) more than two entries on a particular day?
    (*b*) more than four entries in two days?

Interpret the results in (*a*) and (*b*).

Find, furthermore, the probability that

    (*c*) none of the five accounts will receive an entry on a particular day;
    (*d*) more than two accounts will receive at least one entry on a particular day;
    (*e*) exactly two accounts will receive at least one entry on a particular day.

    HINT: the binomial distribution should also be used in (*c*), (*d*), and (*e*), *cf.* worked example in Section 4 above.

10. The Rational Barminster bank has six branches in the centre of a large city. On average, each branch has four customers per day requesting withdrawal of large amounts of cash each exceeding £500. The requests may be assumed to occur at random.

    (*a*) What is the probability of at least five of the branches having four or more such requests on a particular day?

    (*b*) Each branch manager is instructed by head office to keep £2000 aside in cash each day to meet these demands. Do you think this cash reserve is adequate? Justify your answer. What is the expected value of withdrawals from the reserve each day in a particular branch?

State clearly all the assumptions made in answering (*a*) and (*b*).

    (*c*) Had the average number of withdrawals per day varied from branch to branch, would the method of analysis used in (*a*) be adequate? Explain.

11. A factory uses a large number of machines of the same type. In any three-month period the average number of breakdowns due to a particular type of fault is eight. The cost to the company of such a breakdown is estimated at £80. Estimate the number of months in a year in which the company will lose more than £200 due to faulty machinery.

12. A car-hire firm has two cars, which it hires out by the day. The number of demands for a car each day has a Poisson distribution with mean 1·5, and the charge for hire is a flat rate (irrespective of mileage) of £4 per day. Determine the average daily receipts of the firm.

6. The Poisson distribution as an approximation to the binomial distribution.

In discussing the binomial distribution in Section 2 above we deliberately considered examples where the number of trials was comparatively small. However, if n is large the calculation of binomial probabilities is very tedious. With $n = 60$ and a probability $p = 0·1$ of success at each trial, the exact probability of just five successes is:

$$P(5) = \frac{60!}{5! \, 55!} (0·1)^5 \, (0·9)^{55}.$$

To obtain this value (0·166207) is clearly not a trivial matter.

Fortunately in a case such as this an analogy can be drawn between the binomial and Poisson distributions. Admittedly a Poisson random variable can assume any one of an infinite number of values 0, 1, . . ., whereas a binomial variable is limited to the values 0, 1, . . . n. However, when n is large and p is small, the probability of obtaining anywhere near n successes is going to be negligible. As an approximation, then, the assumption that the number of trials is unlimited is quite reasonable, since to obtain a large number of successes is so unlikely. If a success is now regarded as a randomly occurring event the analogy is complete. All that is needed is to determine which Poisson distribution provides the appropriate approximation of a particular binomial distribution. In the example here, the expected number of successes is $np = 60 \times 0·1 = 6$.

Nothing could now be more natural than to use as an approximation the Poisson distribution with this expected value, *i.e.* equal to that of the exact binomial distribution.

Accordingly,

$$P(5) \doteqdot \frac{e^{-6} \, 6^5}{5!}$$
$$= 0·16064.$$

This expression is considerably easier to evaluate than the exact binomial probability, and is an error by only about 3%. Intuitively, one might expect that the larger the value of n, and the smaller the value of p, the better the approximation would be, since the analogy between the distributions is then closer. This is certainly true in general. If $n = 60$ and p is reduced to 0·05, for example, the exact binomial probability of five successes is 0·01016, whilst the Poisson approximation is

$$P(5) = \frac{e^{-3}\,3^5}{5!} = 0{\cdot}10082.$$

Here the error is less than 0·8 %.

Although the approximation is reasonably good for values of n as small as 20, and values of p as large as 0·1, tables of the binomial distribution have been prepared for parameters of this order of size. Its real importance rests on its applicability for very large n and small p when no such tables are available.

The binomial distribution and its Poisson approximation are compared in Fig. 37 for $np = \lambda t = 2$.

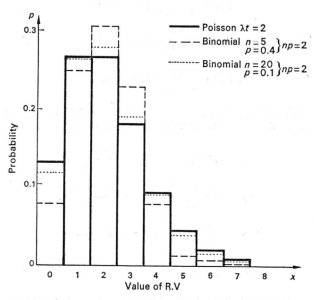

FIG. 37.—Comparison of binomial and Poisson distributions.

Worked Examples

1. An automobile manufacturing company purchases finished headlights from another manufacturer. The components are delivered in batches of 1000, and a batch is considered by the purchaser to be unacceptable if it contains more than 60 faulty components. The purchaser's quality-control department adopts the policy of checking 40 components chosen randomly

from a batch of 1000. If more than 3 of these are faulty the batch is rejected. Analyse the procedure.

Answer

Suppose a batch has 60 faulty components. The probability that a faulty component is selected for examination is $\frac{60}{1000} = 0.06$. (Strictly speaking, this probability changes as the sample is drawn. For example, when we come to the selection of the 40th, if 3 faulty components have already been drawn, the probability that the 40th is faulty will be 57/961. However, as a very good approximation we assume the probability to be constant.) In the selection of 40 (*i.e.*, 40 "trials") the exact probabilities of the indicated number of faulty components are:

$$P(0) = \frac{40!}{0!\,40!}\,(0.06)^0\,(0.94)^{40}$$

$$P(1) = \frac{40!}{1!\,39!}\,(0.06)^1\,(0.94)^{39}$$

$$P(2) = \frac{40!}{2!\,38!}\,(0.06)^2\,(0.94)^{38}$$

$$P(3) = \frac{40!}{3!\,37!}\,(0.06)^3\,(0.94)^{37}.$$

Using a Poisson approximation with mean value of 2·4 (since the expected number of faulty items in the sample is $40 \times 0.06 = 2.4$), we have:

$$P(0) \doteqdot e^{-2.4}, \qquad P(1) \doteqdot \frac{e^{-2.4}\,(2.4)}{1!},$$

$$P(2) \doteqdot \frac{e^{-2.4}\,(2.4)^2}{2!}, \quad P(3) \doteqdot \frac{e^{-2.4}\,(2.4)^3}{3!}.$$

Thus

$$P(\leqslant 3) = e^{-2.4}\,(1 + 2.4 + 2.88 + 2.30)$$
$$= 0.7786.$$

So,

$$P(>3) = 0.2214.$$

We conclude, then, that the probability of rejecting a batch which is just unacceptable in quality is only 0·2214. In other words nearly 78 % of batches with exactly 60 faulty components will be accepted! Clearly this figure will be better (*i.e.*, less) for batches with more than 60 faulty items. However, the procedure used certainly does not appear to be very sound.

2. An electronic assembly consists of 1000 components. Each component is manufactured to high specifications, and will operate correctly with a probability of 0·99985. Failure of any component causes failure of the assembly. What percentage of assemblies will not operate correctly?

Answer

Exactly, P(failure of assembly)

$$= P\text{(failure of one or more components)}$$
$$= 1 - P\text{(failure of no components)}$$
$$= 1 - \frac{1000!}{0!\,1000!}\,(0 \cdot 00015)^0\,(0 \cdot 99985)^{1000}.$$

However, as an approximation, we use the Poisson distribution with mean value $0 \cdot 00015 \times 1000 = 0 \cdot 15$, since there are on average $0 \cdot 15$ faulty components/assembly. Thus

$$P\text{(failure of assembly)} \doteqdot 1 - P(0)$$
$$= 1 - e^{-0 \cdot 15} = 0 \cdot 1393.$$

Even with a high degree of individual component reliability the large number of components in an assembly leads to nearly 14% of the assemblies being faulty.

Exercises on Section 6

13. A life insurance company has found that the probability is $0 \cdot 0001$ that a person in the 50 to 60 age-group will die from a rare disease during a one-year period. If the company has 10,000 policy-holders in this age-group, what is the probability that the company must pay off more than 4 claims because of death from this cause in the coming year? Assume that the disease is neither contagious, nor infectious.

14. A company wishes to reduce the costs of accepting a sub-standard shipment of goods. It is decided to accept a shipment only if there is at the most one defective item in a random sample of 50 taken out of the consignment. Find the probability that the shipment is accepted if it is

    (*a*) $5 \cdot 0\%$ defective;
    (*b*) $2 \cdot 5\%$ defective;

15. Flaws in a certain type of woollen material occur at random with an average of 1 per 1000 square feet. Already 60 rolls have been manufactured, each containing 500 square feet of material. A roll containing more than 2 flaws is sold at a substantially reduced cost. Find

    (*a*) the probability that any particular roll has more than 2 flaws;
    (*b*) the probability that at least 4 of the 60 rolls will have to be reduced in price.

16. An insurance company estimates that 5% of the passengers at an airport purchase flight insurance. If the flow of passengers out of the airport per day is 8000, how many days in the year would you estimate that more than 500 people would purchase insurance? (Assume 1 year = 350 operating days.)

CHAPTER 6

CONTINUOUS PROBABILITY MODELS

INTRODUCTION

In this chapter we turn our attention to the continuous distributions which appear most frequently in statistical work. The list is certainly not intended to be complete, and indeed one or two other continuous distributions will be introduced in later chapters. Much more attention is devoted in the following pages to the normal distribution than to any of the others. The reason for this lies in the fact that much of our later work is based on normally distributed random variables.

THE EXPONENTIAL DISTRIBUTION

1. General. The exponential probability distribution arises most naturally from considerations of the time (or space) elapsing between randomly occurring events. Suppose that events have a Poisson distribution with average rate of occurrence λ. Further suppose that an event has just taken place, and that X is the random variable "time to next event." Now

$$P(X > x) = P(\text{no event occurs in interval } 0, x)$$
$$= e^{-\lambda x} (\lambda x)^0 / 0!$$
$$= e^{-\lambda x}.$$

Since $X > x$ and $X \leqslant x$ are complementary events,

$$P(X \leqslant x) = 1 - e^{-\lambda x}.$$

This expression gives the cumulative probability distribution function (c.d.f.) of X, *i.e.* the area underneath the curve of the probability density function (p.d.f.) between 0 and x. Consequently, the probability density function itself, $p(x)$, is obtained by differentiation:

$$p(x) = \frac{d}{dx} \{1 - e^{-\lambda x}\}$$
$$= \lambda e^{-\lambda x}.$$

This, then, defines the continuous distribution of an exponentially distributed random variable. It is clear from the definition that x can assume any value greater than 0, since there is no theoretical limit to the interval which may elapse before another event occurs. That it is a legitimate distribution can be seen by letting x tend to infinity in the expression for $P(X \leq x)$, and obtaining a value of unity for the total area underneath the curve of the p.d.f. The distribution has the form shown in Fig. 38.

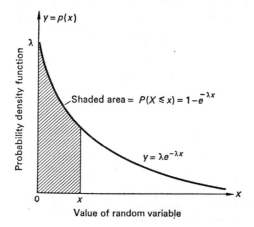

FIG. 38.—Exponential distribution.

The similarity between the profiles of the geometric and exponential distributions is more than coincidental. In fact the latter can be interpreted as a continuous analogue of the former. As such, it could be argued to give a sounder theoretical model of, for example, service-lives (which are, after all, really continuous random variables) than the geometric distribution. However, since from an accounting point of view lives are usually measured in integral numbers of years, a discrete model is perhaps more convenient.

Worked Example

Changes in Bank Rate occur randomly, and on average twice per year. A change has just occurred. What is the probability of a further change within the next 6 months?

Answer

Here
$$\lambda = 2, \ x = \tfrac{1}{2}.$$

Thus
$$\begin{aligned}
P(X \leq \tfrac{1}{2}) &= 1 - e^{-2 \cdot \frac{1}{2}} \\
&= 1 - e^{-1} \\
&= 1 - 0 \cdot 37 \\
&= 0 \cdot 63.
\end{aligned}$$

Note that since λ is given as a rate per year, the interval must be taken as $\tfrac{1}{2}$ (year), not 6 (months).

2. Expected value and variance of exponential distribution. It is interesting to consider the expected value of the exponential distribution in the context of the Poisson distribution from which it may be derived. The parameter λ was defined as the average rate of occurrence of events. The average interval between events is therefore $1/\lambda$. If λ is 2 per year, the average value of X is $\frac{1}{2}$ year = 6 months.

Thus $$E(X) = 1/\lambda.*$$

To find the variance of an exponentially distributed random variable is rather more difficult. We simply quote the result:

$$\text{var } X = 1/\lambda^2.$$

Exercises on Section 2

1. Meetings of a local association of accountants take place randomly, and on average 4 times per year. If a meeting has just been held, what are the chances that another will take place within the coming 4 months?

2. Rises in the retail price of automobiles produced by the Cheetah company occur on average $2\frac{1}{2}$ times per year. Immediately after a price rise has been declared a prospective customer orders a new car for delivery 5 months hence. What is the probability that he will have to bear another price rise when taking delivery of his car?

3. In the illustrative worked example on p. 117, find the probability that

(*a*) Bank Rate does not change in the next year.

(*b*) Bank Rate changes for the first time after between 3 and 5 months time has elapsed.

4. A computer operates without fault on average for 50 hours. Find the probability that it will operate effectively for

(*a*) more than 60 hours;
(*b*) between 40 and 50 hours;
(*c*) less than 42 hours.

5. In the continuous formation of flat glass flaws appear, on average, once every 8 yards. Ten-yard lengths without flaws are sold at full price, whereas

---

* More formally, by definition of expected value,

$$E(X) = \int_0^\infty x\lambda\, e^{-\lambda x}\, dx = \lambda\left[-\frac{x}{\lambda} e^{-\lambda x} - \frac{1}{\lambda^2} e^{-\lambda x} \right]_0^\infty$$

(as a check differentiate [ ] to obtain $x\, e^{-\lambda x}$)

$$= \lambda\left[(0 - 0) - \left(0 - \frac{1}{\lambda^2} \right) \right] = 1/\lambda.$$

In calculating the variance, $E(X^2)$ is found similarly.

ten-yard lengths with flaws, or shorter lengths without flaws have to be reduced substantially in price (or reprocessed).

Given that a flaw has just been detected, what is the probability of obtaining a full price sheet? Estimate the percentage of total production which has to be sold at a reduced price (or reprocessed).

BETA DISTRIBUTION

3. General. A continuous probability model with a finite range is sometimes required, and a beta distribution is often appropriate. The probability density function of a random variable, X, with a beta distribution is defined by:

$$p(x) = c(x - x_1)^m (x_2 - x)^n \qquad x_1 < x < x_2.$$

Here m and n are constants which can have any values greater than zero. The distribution lies between the values $x = x_1$, and $x = x_2$ and is in general negatively skewed, positively skewed, or symmetrical, depending on whether $m > n$, $m < n$, or $m = n$, respectively. The distribution is shown in Fig. 39.

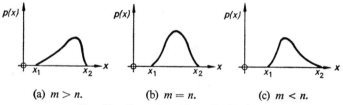

(a) $m > n$.   (b) $m = n$.   (c) $m < n$.

FIG. 39.—Examples of beta distribution.

c is a constant whose value is determined by x_1, x_2, m, n, and whose function is simply to ensure that the area under the curves in Fig. 39 is unity (*i.e.*, that the curves represent true probability distributions). In fact

$$c = \frac{(m + n + 1)!}{(x_2 - x_1)^{m+n+1} m! \, n!}.$$

It is often useful to interpret x_1 and x_2 as the most "pessimistic" (lowest) and "optimistic" (highest) values of the random variable. These two figures, together with a further subjective estimate of the "most likely" value (*i.e.*, the mode) enable a beta distribution with low values of m and n to be used as a model for the behaviour of the random variable.

Finally, the expected value, variance, and mode of a beta distribution are given by

$$E(X) = \frac{(m+1)x_2 + (n+1)x_1}{(m+n+2)}. \qquad (i)$$

$$\text{var } X = \frac{(x_2 - x_1)^2 (m+1)(n+1)}{(m+n+2)^2 (m+n+3)}. \qquad (ii)$$

$$\text{Mode} = \frac{(mx_2 + nx_1)}{(m+n)}. \qquad (iii)$$

Worked Example

Prophet Ltd. have arrived at pessimistic, most likely, and optimistic estimates for next year's revenue of £400,000, £640,000, and £800,000 respectively. Estimate the probability distribution of the revenue.

Answer

Using units of £100,000, we have

$$x_1 = 4, \ x_2 = 8, \ \text{mode} = 6 \cdot 4.$$

A suitable pair of low values of m and n which, together with the above three values, satisfies (iii) are $m = 3$, $n = 2$.

Thus: $$p(x) = c(x - 4)^3 (8 - x)^2$$

where $$c = \frac{6!}{(8 - 4)^6 \, 2! \, 3!}$$

            $$= 0 \cdot 0146.$$

A sketch of this curve on graph paper enables areas under the curve—and therefore probabilities of the revenue falling within any ranges—to be estimated.

Exercises on Section 3

6. The net present value (NPV) of a capital investment project is estimated pessimistically at £3000, optimistically at £6000, with a most likely value of £4000. Suggest a suitable model for the distribution of the NPV, and use your model to find the probability that it exceeds £4500. Find further the expectation of the NPV, and use your result to estimate the combined NPV of 15 very similar projects undertaken simultaneously.

7. A department's annual expenditure is known not to have exceeded £16,000 nor to have been less than £11,000 over the past few years. On average the annual expenditure was in fact £14,000, with a standard deviation of £1000. Substitute these values in the expressions for the expectation and variance of a beta distribution, and verify that $m = 2$, and $n = 1$.

Hence find the probability that departmental expenditure will exceed £13,000 next year. (All figures given are in terms of present values.)

8. The inventory level of a firm is never allowed to fall below £30,000, and it is not thought economic to maintain stocks at more than £50,000.

The most frequent level is £43,000. Use an appropriate simple beta distribution model to estimate the mean stock level. Determine the proportion of time that the inventory value falls below £35,000 in your model. All figures may be assumed to be present values.

9. The salvage value of an asset in three years' time is unknown. Analysis of data on similar three-year-old assets in the past reveals that—in terms of present values—salvage value was never less than £200, nor more than £500. In addition, the mean salvage value was £300 with a standard deviation of £37·80. Verify that a Beta distribution with $m = 1$, $n = 3$, provides a suitable model, and hence estimate the number out of a group of 80 newly acquired assets that are likely to have a salvage value less than £350 in three years' time.

THE NORMAL DISTRIBUTION

4. Introduction. There is no doubt that the most frequently used (and indeed, abused!) distribution is the so-called *normal*.

Its properties were first investigated in detail by a German mathematician, Gauss (hence the alternative name, Gaussian distribution) in a scientific context. If an experiment to measure, say, the electric current in a wire was repeated over and over again, it had been observed that the general shape of the histogram of results was always of the shape shown in Fig. 40.

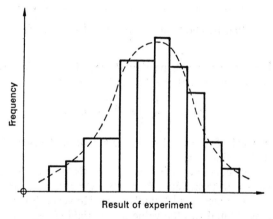

FIG. 40.—Histogram of experimental results.

Verbally, results tended to be symmetrically distributed about a central value, with increasing rarity of occurrence as distance above or below this central value increased.

Gauss was able to derive the theoretical probability density function of the random variable representing the measurement made in the experiment, by assuming that the differences in measured results from one experiment to another were due to a very large number of independent sources, such as changes in temperature, pressure, humidity, and so on.

The simplest type of normal curve has a probability density function $p(x)$ given by:

$$p(x) = \frac{1}{\sqrt{(2\pi)}} e^{-\frac{1}{2}x^2} \qquad -\infty < x < \infty.$$

The graph of $p(x)$ appears in Fig. 41.

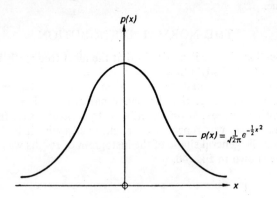

FIG. 41.—Probability density function of $N(0, 1)$.

Several points are worth particular attention at this stage. Firstly the curve is symmetrical about the vertical axis, since

$$p(x) = \frac{1}{\sqrt{(2\pi)}} e^{-\frac{1}{2}x^2} = p(-x).$$

In other words, for any value, x, the height of the curve is the same as it is for the value $-x$.

Secondly, the curve never reaches the x-axis however large the value of x (positively or negatively). A random variable that can in practice assume infinite values is, admittedly, rare. However, it may be difficult to place exact bounds on the value of a variable (*e.g.*, profits, or total claims on an insurance company), and a theoretical distribution which has no bounds often provides a useful model. Finally, the appearance of the factor $1/\sqrt{(2\pi)}$ in the p.d.f. requires some explanation. Remember that for $p(x)$ to be a p.d.f., the area between the curve and the x-axis from $-\infty$ to $+\infty$ must be unity. It so happens that

$$\int_{-\infty}^{\infty} e^{-\frac{1}{2}x^2} \, dx = \sqrt{(2\pi)}.$$

Consequently division of $e^{-\frac{1}{2}x^2}$ by a scale factor of $\sqrt{(2\pi)}$ gives a function which can legitimately fulfil the role of a p.d.f.

5. The standard normal variable: expectation, variance, and probability statements.

A random variable, X, with a p.d.f. given by

$$p(x) = \frac{1}{\sqrt{(2\pi)}} e^{-\frac{1}{2}x^2}$$

is known as a *standard normal variable*. From symmetry considerations alone, the expected value of such a variable is zero. It can also be shown that the variance is one.*

The statement that "the random variable, X, has a standard normal distribution" (*i.e.*, a normal distribution with expected value 0 and variance 1) is summarised by the shorthand notation:

$$X \sim N(0, 1)$$

As with any continuous random variable the areas underneath the curve of the p.d.f. represent probabilities, and to calculate any particular area it is necessary to integrate.

Thus,

$$P(1 \leqslant x \leqslant 1\cdot5) = \int_{1\cdot0}^{1\cdot5} \frac{1}{\sqrt{(2\pi)}} e^{-\frac{1}{2}x^2} \, dx.$$

Extensive tables have been prepared for the purposes of evaluating probabilities like this, and a short version of these tables is included in Appendix II.

6. Use of normal tables.

Tables of the standard normal distribution vary slightly in their presentation. In this book values of the function $\Phi(x)$ (area underneath the curve from $-\infty$ to x) are readily found for values of $x \geqslant 0$. A graphical interpretation of $\Phi(x)$, which is of course the cumulative probability distribution function, is given in Fig. 42.

$$* \; \mathrm{E}(X) = \int_{-\infty}^{\infty} \frac{x}{\sqrt{(2\pi)}} e^{-\frac{1}{2}x^2} \, dx = -\frac{1}{\sqrt{(2\pi)}} \left[e^{-\frac{1}{2}x^2} \right]_{-\infty}^{\infty}$$

$$= \frac{1}{\sqrt{(2\pi)}} [0 - 0] = 0.$$

Similarly,

$$\mathrm{E}(X^2) = 1$$

and

$$\mathrm{var}\, X = \mathrm{E}(X^2) - [\mathrm{E}(X)]^2$$
$$= 1.$$

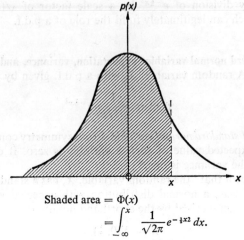

Shaded area $= \Phi(x)$

$$= \int_{-\infty}^{x} \frac{1}{\sqrt{2\pi}} e^{-\frac{1}{2}x^2} \, dx.$$

FIG. 42.—Probability distribution function, $\Phi(x)$, of $N(0, 1)$.

Since the curve is symmetrical, the area to the left of $x = 0$ is 0·50. Consequently for $x > 0$, $\Phi(x) > 0·50$. The symmetry of the curve also makes it unnecessary to tabulate $\Phi(x)$ for negative values of x. Referring to Fig. 43, the two shaded areas are obviously equal. Now the area to the left of $-x$ is, by definition, $\Phi(-x)$. Furthermore, the area to the left of x is, again by definition, $\Phi(x)$, implying that the area to the right of x is $1 - \Phi(x)$ (since the total area under the curve is unity).

Thus, $\Phi(-x) = 1 - \Phi(x)$.

We are now in a position to use the tables to find the probability of x lying in any desired range (x_1, x_2). Three slightly different situations can arise, and each is considered in turn, together with an appropriate example.

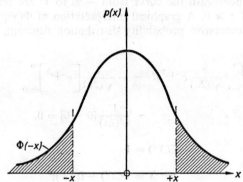

FIG. 43.—Distribution function of $N(0, 1)$ for negative values of x.

(a) x_1, x_2 both positive (*see* Fig. 44 (a)).

Since $\Phi(x_2)$ is the area from $-\infty$ to x_2, and $\Phi(x_1)$ that from $-\infty$ to x_1, the area standing on the range (x_1, x_2) is $\Phi(x_2) - \Phi(x_1)$.

Therefore,

$$P(x_1 \leqslant x \leqslant x_2) = \Phi(x_2) - \Phi(x_1)$$

e.g.,

$$\begin{aligned}
x_1 &= 0\cdot6 & \Phi(0\cdot6) &= 0\cdot726 \\
x_2 &= 1\cdot9 & \Phi(1\cdot9) &= 0\cdot971 \\
P(0\cdot6 &\leqslant x \leqslant 1\cdot9) &= 0\cdot971 &- 0\cdot726 \\
& & &= 0\cdot245.
\end{aligned}$$

(b) x_1 negative, x_2 positive (*see* Fig. 44 (b)).

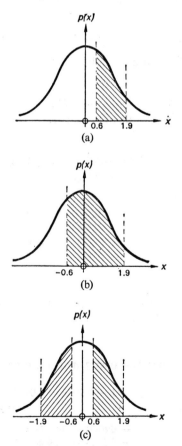

FIG. 44.—Illustration of probabilities associated with a standard normal variable.

The equation

$$P(x_1 \leqslant x \leqslant x_2) = \Phi(x_2) - \Phi(x_1)$$

is still appropriate. However, the value of $\Phi(x_1)$ cannot be found immediately from the tables,

e.g.,

$$\begin{aligned} x_1 &= -0\cdot6 & \Phi(0\cdot6) &= 0\cdot726 \\ x_2 &= 0\cdot9 & \Phi(0\cdot9) &= 0\cdot971 \end{aligned}$$

$$\begin{aligned} P(-0\cdot6 \leqslant x \leqslant 1\cdot9) &= \Phi(1\cdot9) - \Phi(-0\cdot6) \\ &= \Phi(1\cdot9) - \{1 - \Phi(0\cdot6)\} \\ &= 0\cdot971 - 0\cdot274 \\ &= 0\cdot697. \end{aligned}$$

(c) x_1, x_2 both negative (*see* Fig. 44 (*c*)).

Although

$$P(x_1 \leqslant x \leqslant x_2) = \Phi(x_2) - \Phi(x_1)$$

both $\Phi(x_1)$ and $\Phi(x_2)$ are unobtainable directly from the tables,

e.g.,

$$\begin{aligned} x_2 &= -0\cdot6 & \Phi(0\cdot6) &= 0\cdot726 \\ x_1 &= -1\cdot9 & \Phi(1\cdot9) &= 0\cdot971. \end{aligned}$$

By symmetry,

$$\begin{aligned} P(-1\cdot9 \leqslant x \leqslant -0\cdot6) &= P(0\cdot6 \leqslant x \leqslant 1\cdot9) \\ &= 0\cdot245. \end{aligned}$$

Alternatively,

$$\begin{aligned} P(-1\cdot9 \leqslant x \leqslant -0\cdot6) &= \Phi(-0\cdot6) - \Phi(-1\cdot9) \\ &= \{1 - \Phi(0\cdot6)\} - \{1 - \Phi(1\cdot9)\} \\ &= 0\cdot274 - 0\cdot029 \\ &= 0\cdot245. \end{aligned}$$

Worked Example

Find the probability that a random variable with a standard normal distribution has a value between $-1\cdot3$ and $+0\cdot2$.

Answer

From the tables

$$\Phi(1\cdot3) = 0\cdot9032, \quad \Phi(0\cdot2) = 0\cdot5793.$$

Thus

$$\Phi(-1\cdot3) = 1 - 0\cdot9032 = 0\cdot0968.$$

The required probability

$$\begin{aligned} &= \Phi(0\cdot2) - \Phi(-1\cdot3) \\ &= 0\cdot5973 - 0\cdot0968 \\ &= 0\cdot5005. \end{aligned}$$

Exercises on Section 6

10. Find the probability that a random variable with a standard normal distribution lies between

    (*a*) 0·45 and 0·77;
    (*b*) −0·29 and −0·11;
    (*c*) −0·87 and 1·13.

7. The general normal distribution. It is useful before looking at the more general normal model to consider the probability evaluations in Section 6 in a slightly different light. Recalling that the expected value or mean of the standard normal distribution is zero, and that the variance (and therefore, standard deviation) is unity, one can interpret statements like

$$P(0·6 \leqslant x \leqslant 1·9), \text{ or } P(−0·6 \leqslant x \leqslant 1·9),$$

as the probability that *x* lies between 0·6 and 1·9 standard deviations away from the mean, or between −0·6 and 1·9 standard deviations away from the mean.

More generally, a random variable, *X*, with a density function defined by

$$p(x) = \frac{1}{\sigma\sqrt{(2\pi)}} e^{-\frac{1}{2}\left(\frac{x-\lambda}{\sigma}\right)^2} \qquad -\infty < x < \infty$$

is known as a normal distribution with expectation μ and variance σ^2 (*i.e.*, standard deviation σ). In a natural extension of the notation previously introduced, this information is summarised by

$$X \sim N(\mu, \sigma^2).$$

The shape of the distribution for all values of μ and σ is very similar to that of the standard normal distribution. Essentially, the only differences are that the distribution is displaced from the origin by an amount μ, and is "flatter" (if $\sigma^2 > 1$) or more sharply "peaked" (if $\sigma^2 < 1$). (*See* Fig. 45.)

It would be extremely tedious and inefficient if a separate set of probability tables had to be consulted for every different combination of μ and σ. Fortunately this is not the case, and by using a process of *standardisation* any probability calculation involving a normal distribution can be reduced to one involving tabulated standard normal values.

Suppose that the daily takings *X* of the millinery department in a store have, in the past, had an average value of £350 with a standard deviation of £90. An appropriate theoretical model of the distribution of takings is the normal one, with expectation £350 and standard

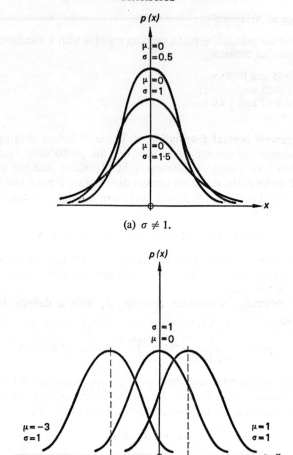

(a) $\sigma \neq 1$.

(b) $\mu \neq 0$.

Fig. 45.—Distributions of normal random variables.

deviation $\sigma = £90$. Graphically the appearance of the distribution is shown in Fig. 46.

Admittedly X is, strictly speaking, a discrete variable (with intervals of $\frac{1}{2}$p between successive possible values), and cannot be negative. The distribution $N(350, 8100)$ fails on both these counts. Firstly, however, the intervals of $\frac{1}{2}$p are sufficiently small, in comparison with a typical day's takings, for the assumption of continuity to be a very good approximation.

Secondly, the probability in the theoretical model of obtaining a value less than zero is extremely small, and therefore unlikely to make

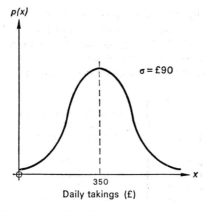

p(x)

$\sigma = £90$

350
Daily takings (£)

FIG. 46.—Daily takings in a millinery store.

any material difference to any conclusions that might be drawn from the model.

Having chosen the theoretical distribution, the problem is now to evaluate the probability that, say, X lies between £370 and £420. The procedure is quite simple. Firstly, we find the deviations of each end of the range from the expected value. These are £370 − £350 = £20 and £420 − £350 = £70 respectively. Secondly, we express each of these in terms of numbers of standard deviations. This provides 20/90 = 0·222 and 70/90 = 0·778, as the number of standard deviations away from the mean of the endpoints of the range. Now referring to standard normal tables in the usual way,

$$P(£370 \leqslant X \leqslant £420) = \Phi(0·778) - \Phi(0·222)$$
$$= 0·7814 - 0·5879$$
$$= 0·1935.$$

Similarly, the probability that X lies between £300 and £400 is found as:

$$P(£300 \leqslant X \leqslant £400) = \Phi\left(\frac{400 - 350}{90}\right) - \Phi\left(\frac{300 - 350}{90}\right)$$
$$= \Phi(0·556) - \Phi(-0·556)$$
$$= 0·4218.$$

Two points in standardising variables should be emphasised:

(a) The mean, μ, is always subtracted from the value to be standardised.

(b) Division is by the S.D., σ, and not the variance, σ^2.

The basis of this procedure rests on the remarkable fact that all normal distributions have the same proportions of their total areas between the same limits, as long as these limits are measured in terms of standard deviations from the mean.

Three particularly useful proportions are given below:

Interval	Proportion of total area under any normal curve within specified interval (%)
$\mu \pm \sigma$	68·3
$\mu \pm 2\sigma$	95·5
$\mu \pm 3\sigma$	99·7

If three random variables have distributions $N(3, 9)$, $N(4, 4)$, and $N(0, 16)$ respectively then 68·3% of the total area lies between 3 ± 3 (*i.e.*, 0 and 6) in the first case, between 4 ± 2 (*i.e.*, 2 and 6) in the second, and 0 ± 4 (*i.e.*, -4 and 4) in the third.

Worked Examples

1. The average number of photocopies made per day in a certain office is 356 with a standard deviation of 55. It costs the company 3p per copy. Determine the probability that the cost on a particular day is

    (a) between £8 and £10;
    (b) less than £11;
    (c) greater than £13.

Answer

    (a) A cost on a particular day of between £8 and £10 implies that between 267 and 333 copies are made. Let C, N be the random variables, "cost" and "number of copies" respectively.

Thus,

$$P(\pounds 8 < C < \pounds 10) = P(267 \leqslant N \leqslant 333)$$
$$= \Phi\left(\frac{333 - 356}{55}\right) - \Phi\left(\frac{267 - 356}{55}\right)$$
$$= \Phi(-0{\cdot}4182) - \Phi(-1{\cdot}6182)$$
$$= 0{\cdot}2846.$$

    (b)
$$P(C < \pounds 11) = P(N \leqslant 366)$$
$$= \Phi\left(\frac{366 - 356}{55}\right)$$
$$= \Phi(1{\cdot}8182)$$
$$= 0{\cdot}9656.$$

    (c)
$$P(C > \pounds 13) = 1 - P(C \leqslant \pounds 13)$$
$$= 1 - P(N \leqslant 433) = 1 - \Phi(1{\cdot}4)$$
$$= 0{\cdot}0808.$$

2. The weights of 1 kilogram packets of sugar may be assumed to be approximately normally distributed with a standard deviation of 3 grams. The filling machine is set to deliver a mean weight, μ, of sugar such that only 1% of the packets have a weight below 1 kilogram. The weekly output is 500,000 packets, and the cost, C, in pence, of producing a packet containing w grams is

$$C = 1\cdot25 + 0\cdot0065w$$

(a) Determine the average cost per week.

(b) If, by installing new machinery, the standard deviation could be reduced to 0·6 grams, and the mean weight altered to give the same percentage below 1 kilogram as before, determine the average savings in £ per week.

Answer

(a) The average cost per packet is the average value (or expected value) of C. This is given by

$$1\cdot25 + 0\cdot0065\mu$$

where μ is the expected value of the weight, w.

We require then to find the expected value, μ, of the normal distribution which has a standard deviation of 3 grams, and which has the property that only 1% of the area is below 1 kilogram (*i.e.*, 1000 grams). Thus

$$\Phi\left(\frac{1000 - \mu}{3}\right) = 0\cdot01.$$

From the tables of the standard normal distribution

$$\Phi(-2\cdot33) = 0\cdot01$$
$$\therefore \quad (1000 - \mu)/3 = -2\cdot33$$
$$\mu = 1007.$$

Finally, the average cost per packet

$$= 1\cdot25 + (0\cdot0065 \times 1007)\text{p}$$
$$\doteqdot 7\cdot79\text{p}.$$

(b) Suppose the standard deviation is now 0·6 grams. Following the same line of argument as before,

$$\Phi\left(\frac{1000 - \mu}{0\cdot6}\right) = 0\cdot01$$

$$(1000 - \mu)/0\cdot6 = -2\cdot33$$
$$\mu = 1001\cdot4.$$

Average cost per week

$$= 1\cdot25 + (0\cdot0065 \times 1001\cdot4)\text{p}$$
$$= 7\cdot76\text{p}.$$

On an output of 500,000 packets the saving is

$$= 500,000 \times 0 \cdot 03p$$
$$= £150.$$

Exercises on Section 7

11. The mean income in a certain area is £4271 with a standard deviation of £1300. Estimate the proportion of the population whose incomes are

    (*a*) between £3600 and £4500;
    (*b*) between £3000 and £4000;
    (*c*) less than £5000;
    (*d*) greater than £2900.

12. The accounts receivable of a company have an average size of £10,000 and a standard deviation of £2000. If an account is selected at random what is the probability that its size lies between £9000 and £12,000?

13. A building contractor is planning his budget for overtime wages over the three-month summer period, and allows for 1000 hours. In the past, the number of overtime hours worked in this quarter of the year has averaged 800 hours with a standard deviation of 100 hours. Determine the probability that

    (*a*) his budget is exceeded;
    (*b*) the number of overtime hours worked will be between 800 and 1000.

14. A manufacturer of hair tonic is considering production of a new hair dressing which he hopes will increase his sales. The incremental profit is £1 per unit, and the necessary investment in equipment is £50,000. Yearly demand is estimated to have a mean value of 48,000 units, with a standard deviation of 5000 units. Estimate the probability that in the first year he will

    (*a*) recover the cost of the new equipment;
    (*b*) fail to recover the cost of the new equipment by more than £3000.

Prices and costs may be assumed to be in terms of present values.

8. The normal curve as an approximation to the binomial distribution. The Poisson distribution has already been suggested as a suitable approximation to the binomial distribution when *n* is large and *p* (or $1 - p$) is small. However, if *p* is not near 0 or 1 the Poisson approximation cannot be used. The binomial distribution illustrated in Fig. 34 has the same inverted bell shape when $p = 0 \cdot 5$ as a normal distribution. Since it can also be seen that this similarity increases as *n* increases, it comes as no surprise to learn that for large values of *n*, and for values of *p* not near 0 or 1, the normal curve can be used as a good approximation to the binomial distribution.

Suppose that, on average, 30% of the transactions recorded by a company involve sums in excess of £100, and that on a particular day

25 transactions are recorded. The probability that just 5 of these transactions involved sums greater than £100 is given exactly by:

$$P(5) = \frac{25!}{20! \, 5!} (0 \cdot 30)^5 (0 \cdot 70)^{20}$$
$$= 53{,}130 \, (0 \cdot 00243)(0 \cdot 0007979)$$
$$= 0 \cdot 1030.$$

The calculation of this result is laborious, however, both in terms of the large factorials involved, and of the high powers.

As an approximation we use the normal distribution with the same mean and variance as the exact binomial distribution. Consequently we set

$\mu = np = 25 \times 0 \cdot 30 = 7 \cdot 5$ and
$\sigma = \sqrt{\{np(1-p)\}} = \sqrt{\{25 \times 0 \cdot 30 \times 0 \cdot 70\}} = 2 \cdot 2913.$

Since the normal distribution is continuous, the probability of *exactly* 5 is zero. However, by taking the range 4·5 to 5·5 and finding the normal probability of a value falling in that range, an approximation to the exact (discrete) binomial probability of 5 can be found:

$$P(4 \cdot 5 < \text{normal variable} < 5 \cdot 5)$$
$$= \Phi\left(\frac{5 \cdot 5 - 7 \cdot 5}{2 \cdot 2913}\right) - \Phi\left(\frac{4 \cdot 5 - 7 \cdot 5}{2 \cdot 2913}\right)$$
$$= \Phi(-0 \cdot 8730) - \Phi(-1 \cdot 3095)$$
$$= 0 \cdot 0963.$$

Comparing this with the exact binomial probability we see that the error incurred in using the approximation is 6·5%.

A value for p nearer to 0·50 results in a better approximation, since the exact binomial distribution is less skewed, and therefore more like the symmetrical normal distribution.

For example, with $p = 0 \cdot 40$:

$$P(5)_{\text{binomial}} = \frac{25!}{20! \, 5!} (0 \cdot 40)^5 (0 \cdot 60)^{20}$$
$$= 53{,}130 \, (0 \cdot 01024)(0 \cdot 00003656)$$
$$= 0 \cdot 0199.$$
$$P(5)_{\text{normal}} = \Phi\left(\frac{5 \cdot 5 - 10}{2 \cdot 449}\right) - \Phi\left(\frac{4 \cdot 5 - 10}{2 \cdot 449}\right)$$
$$= \Phi(-1 \cdot 8374) - \Phi(-2 \cdot 2458)$$
$$= 0 \cdot 0207.$$

The error in this approximation is only 4%.

A useful feature of the normal curve approximation to the binomial distribution is its ability to provide probabilities for ranges. This is illustrated in the following worked example.

Worked Example

In the situation above, calculate the probability that the number of transactions in excess of £100 is greater than 4 and less than 9.

Answer

Using the normal approximation with $\mu = 7.5$ and $\sigma = 2.2913$, and a range from 4·5 to 8·5,

$$P(4 < \text{number of transactions in excess of £100} < 9)$$
$$= \Phi\left(\frac{8.5 - 7.5}{2.2913}\right) - \Phi\left(\frac{4.5 - 7.5}{2.2913}\right)$$
$$= \Phi(0.4365) - \Phi(-1.3095)$$
$$= 0.6687 - 0.0951$$
$$= 0.5736.$$

NOTE: Exactly, the binomial model gives

$$P(4 < \text{number of transactions exceeding £100} < 9)$$
$$= P(5) + P(6) + P(7) + P(8)$$
$$= 0.1030 + 0.1471 + 0.1712 + 0.1651$$
$$= 0.5864.$$

The error is only 2·2%, and the normal approximation does not involve the lengthy separate calculation of the four discrete probabilities.

It is difficult to lay down an exact set of rules for when to use the normal curve approximation, and when not. Table 24, however, provides a rough guide.

Table 24. Use of normal approximation to binomial distribution

Value of p	Use approximation if $n \geqslant$
0·50	10
0·45 or 0·55	12
0·40 or 0·60	13
0·35 or 0·65	15
0·30 or 0·70	17
0·25 or 0·75	20
0·20 or 0·80	25
0·15 or 0·85	34
0·10 or 0·90	50
0·05 or 0·95	100
0·025 or 0·975	200
0·010 or 0·990	500

Exercises on Section 9

15. On average, 20% of the invoices received by a purchasing department require immediate attention, whilst the remaining 80% can be delayed.

Determine the probability that out of a batch of 60 invoices, more than 15 will need to be dealt with immediately.

16. The probability that any particular departmental budget is exceeded is 0·40. Determine the approximate probability that less than six out of a group of 25 budgets will be exceeded. You may assume that departmental expenditures are independent of each other.

17. The probability that an asset will be retired after one year is 0·35. Estimate the probability that, out of a group of 45 identical assets, the number retired after a year is less than 22 and more than 11.

18. A company is faced with the decision of investing in one of three alternative projects, *A*, *B*, and *C*. An analysis of market trends and general economic conditions produces the following information concerning the NPVs of the projects:

Project	Expected NPV (£)	S.D. of NPV (£)
A	50,000	3000
B	45,000	2600
C	52,000	3500

Assuming normal distributions for the NPVs analyse the data further, and present a recommendation to the board of the company, making (and clearly stating) any necessary assumptions.

19. Piscator Limited manufacture a wide range of fish soups. In particular, the demand for the eel soup has a mean of 600 packets per day, and a standard deviation of 200 per day. Unfortunately, because of difficulties of retaining their product at its peak of piscine freshness, over-production results in their having to dispose of the excess at reduced prices. Under-production, furthermore, causes loss of goodwill, loss of sales, and so on. In fact, it is estimated that the costs of over-production are 10p per packet, and of under-production, 8p per packet.

Explain carefully (without calculating its exact value) how you would obtain the most satisfactory level at which to maintain daily production.

20. The weekly sales revenue for a product has a mean of £1237·42, and a standard deviation of £212·14. Give a range in which 90% of weekly sales revenue will fall.

21. Assuming that proposed mergers and takeovers appear before the Monopolies Commission randomly at the rate of three per month (4 weeks), estimate the number of weeks in a year (48 weeks) in which no new cases are considered by the Commission.

22. A supermarket has 15 cash-outs, each being the responsibility of a particular cashier. It is suspected that three of the cashiers are not registering correct sales on the tills, but it is not known which three. Accordingly random checks of four tills at a time are instituted. Determine the probability that in one of these checks

(*a*) all three dishonest cashiers are detected;
(*b*) just two are detected;

    (c) only one is detected;
    (d) none is detected.

23. Assuming that the probability of a salesman making a sale at a randomly selected house is 0·1, and that a salesman makes 20 calls per day, determine the following:

    (a) Probability of no sales.
    (b) Probability of at least one sale.
    (c) Probability of four or more sales.
    (d) Probability of exactly four sales.

24. A supermarket cashier makes at least one error on 15% of the customers' bills at the check-out. What is the probability that in serving four customers she makes no mistakes?

CHAPTER 7

SAMPLING THEORY AND ESTIMATION

INTRODUCTION

Most of us have at some time met the notion of sampling—perhaps through the opinion polls which proliferate before an election, perhaps through completing questionnaires in the course of our work, or even through sipping a little home-made wine or beer before bottling! Not as familiar, however, are the methods by which sample information is collected, and by which conclusions are drawn from it. It is with discussing these methods that we shall be concerned in this chapter.

A very large section of statistical analysis is dependent on the concept of sampling, either explicitly or implicitly. Let us first of all, therefore, be quite clear of the statistical meaning of some of the terms which we shall be using. A *population* is the complete set of items which are of interest to us in our particular investigation. Note that a population need not necessarily consist of people, but may, for example, be of machines of a particular type, all accounts at a Building Society branch, all the values of a random variable, or all the factories in a particular region. Any part of the population is known as a *sample*. The sample, which is therefore a subset of the population, is extracted from the population by the process of *sampling*. Given the required size of sample, there are, of course, many samples of different composition which may be drawn from the same parent population.

Sampling can be invaluable for a number of reasons. In the first place, the act of sampling may necessarily result in the destruction of the sampled item. Obviously the manufacturer of electronic components—say transistors—wishes to ensure that his products are meeting certain standards of quality. In order to do this thoroughly, a transistor should be tested to destruction—an impossibility as far as the manufacturer is concerned since he needs some to sell! He may, however, escape from his dilemma by testing to destruction only a sample from his finished goods, in the hope that the performance of the sampled transistors will be more or less typical of the population. The population in this case is, of course, the totality of transistors that the manufacturer produces.

Secondly, sampling is important from the point of view of economy. It would, for example, be costly to check for correctness every transaction recorded in a computerised system of accounts. However, by selecting for checking only a sample from the population of transactions it is possible to obtain a reasonably accurate picture of the system's performance.

Again, the time element may be an important reason for sampling. In an election campaign it is important to the various parties to know the trend of public opinion. It would, however, be impossible because of the time involved (and, indeed, other reasons) for the opinion of every elector to be obtained before the election. By the end of such an enquiry, public opinion may have altered to an extent that entirely invalidates the results. Clearly, the solution here is to base inferences about electoral behaviour on the results of a judiciously chosen sample of electors.

Finally, sampling may not only be valuable, but in fact essential, when part of the population is unavailable for study. It would be impossible, for instance, for the actuaries and underwriters of an insurance company to obtain details of the cause of every motor accident, since the occupants may be killed, and the vehicle destroyed.

Of course a sample cannot possibly include all the information contained in the population. Any inferences made about the population in the light of the sample are subject to error. This is referred to as the *sampling error*. Fortunately, however, the sampling error can often be made quite small by suitably choosing the sample. Equally importantly, as we shall see later, specific quantitative statements can be made about the limits of the sampling error.

A second source of error in sampling is *bias*. This may arise through a sample being drawn in such a way that it is atypical of the population. Thus an investigation into per capita income, based on a selection of names from a telephone directory, will lead to biased results, since the less wealthy are not telephone subscribers. A final source of error is due neither to sampling, nor to the manner in which the sample is chosen, but simply to the fact that inaccurate information may be collected. This is exemplified by the respondent who does not wish to disclose his annual income, and supplies an incorrect figure!

Having drawn attention to the advantages of sampling, we turn now to discussion of the manner in which samples may be drawn from a population.

SAMPLING SCHEMES

1. Simple random sampling. This method of sampling is based on the assignation of equal probabilities of selection to each item in the population.

Suppose for the sake of argument that a population of accounts has 100 members, and that these are numbered consecutively from

1 to 100. To select a random sample of size 20 all that we need do is to list 20 numbers at random between 1 and 100, and include in our sample the corresponding members of the population. In actual fact it is not at all easy to write down a list of numbers and be sure that they are random. Most people have particular preferences for certain numbers or sequences, and would introduce an involuntary bias into their list. A much more efficient and objective way of generating a list of random numbers is to consult a set of random number tables, examples of which are included in Appendix III. These are prepared by electronic or mechanical, rather than human, means. The Premium Bond draw performed by ERNIE is perhaps the best known example of the use of an electronic random number generator, and is used for drawing a sample (the winning Bond numbers) from a population (the complete set of purchased Bond numbers).

Using the table in Appendix III we commence at any point and read off 20 pairs of numbers in order:

44	22	78	84	26
04	33	46	09	52
68	07	97	06	57
74	25	65	76	59

Notice that in fact our population is better numbered from 00 to 99 rather than from 01 to 100, since we then avoid the complication of having to extract 20 numbers from 99 with 2 digits, and one with 3 digits.

Although simple random sampling is not always the most efficient method of obtaining information about a population, it does form the basis of more complicated schemes, some of which are outlined below.

2. Stratified random sampling. This sampling scheme rests on the division of the parent population into natural *strata* or categories, followed by random sampling from each of these strata (*see* Fig. 47 (*a*)). For example, in a sample audit the items in an inventory may range from the very expensive and rare, to the inexpensive and numerous. A stratified sampling scheme would involve the selection of a random sample from each cost stratum.

In some cases it is useful to choose from each stratum a fixed proportion of the members of that stratum, but this is certainly not mandatory. The sample drawn from each stratum is, however, always a random one (*i.e.*, within a stratum each item is equally likely to be sampled).

3. Multi-stage sampling. This method involves sampling in two or more distinct steps. By way of illustration of a two-stage sampling plan, consider the Rational Barminster Bank which has branches in

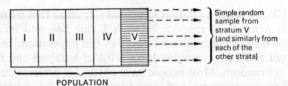

(a) Stratified random sampling (strata are homogeneous sub-populations).

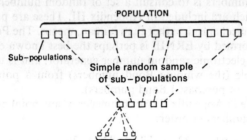

(b) Multi-stage sampling.

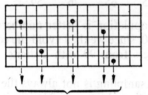

Random sample of sub-populations

(c) Cluster sampling (sub-populations are heterogeneous in content).

FIG. 47.—Sampling schemes.

most towns and cities. An investigation into the bank charges levied might be carried out by firstly taking a simple random sample from all the branches in the country, followed by the selection of a second series of random samples of customer accounts from each of the branches included in the first sample. The procedure is illustrated diagrammatically in Fig. 47 (b).

A number of variations may be made on the same theme, apart from the obvious one of using an increased number of stages in the sampling plan. In our bank example, for instance, at the first stage a stratified scheme could have been used (with strata determined by, say, the sizes of towns in which branches are situated), followed by a set of simple random samples at the second stage. Alternatively, the second stage could involve another stratified scheme with size of account defining the strata.

4. Cluster sampling. This scheme depends on a natural grouping of items in the population to form a set of "sub-populations" or clusters. A random sample of these groups is then drawn to form the final sample (*see* Fig. 47 (*c*)). In our previous illustration the branches of the bank may be grouped initially on a geographical basis to form 500 regions, say, covering the entire country. From the 500 regions a simple random sample of regions would be drawn, and all the branches in each of the selected regions investigated.

It is interesting to note, incidentally, that a stratified sampling scheme is most efficient when the items within each stratum are homogeneous, and the strata are as different as possible from each other. A cluster sampling procedure, on the other hand, is most efficient when the items within the cluster are heterogeneous, and the differences between clusters as small as possible.

5. Systematic sampling. This provides a very convenient way of selection, since if a 10% sample is required we simply take every tenth item of the population. Although samples obtained in this way may be more or less random, systematic sampling can have serious defects if an unknown cyclical factor exists in the population. To test manufactured components on this basis will result in a severely biased result if the machinery develops a fault—such as a damaged cog on a toothed wheel—which causes every tenth item to be defective!

6. Judgment and convenience sampling. Finally two schemes should be mentioned, both of which lack objectivity in their mode of selection.

Judgment sampling depends on the opinion of the investigator as to which items or groups of items to sample. Thus an auditor may decide that it is only necessary to examine certain categories of account. Although the method may sacrifice fairness of representation in the interests of economy (of time and/or cost), it can certainly have value in a preliminary or pilot study, where an expert view may avoid the needless sampling of items irrelevant to the study.

Convenience sampling involves the examination of some of the most readily accessible items in the population. This has obvious disadvantages with regard to bias, but may be justified if only approximate results are required.

SAMPLING OBJECTIVES

Whichever method of sampling is most suited to a specific investigation depends on a number of factors, including the size of sample which can be effectively studied, and the objectives of sampling. It is impossible to discuss these matters fully in this book, and the interested reader is advised to consult the references in the Bibliography, pp. 297–8 for further detail. In particular Ijiri and Kaplan provide a valuable contribution to the literature on sampling in auditing. The most important purposes of sampling are mentioned briefly below.

7. Acceptance sampling. In the case of any form of quality control, either of manufactured products, or of, say, the accuracy of an invoicing system, the objective of sampling is to accept (or reject) a batch (population) of items as being suitable (or unsuitable). Thus, a sample of the punched cards on which a company records for each of its employees the number of hours worked, the various rates of pay, and deductions for tax and other purposes may be taken. On the basis of the number of cards in the sample with incorrect punching the total population of cards may be accepted or rejected. The critical number of erroneous cards in the sample, which would trigger the decision to reject, depends in a precise way on the number of incorrect cards in the population, on the sample size, and on the margin for error (*i.e.*, probability of making a wrong decision) which is considered permissible.

8. Discovery sampling. A second objective of sampling may be the detection of, for example, fraud. Obviously nothing less than a complete examination of all the records would enable an auditor to say definitely that no fraud had been committed. It is possible, however, to determine the size of sample necessary for us to be able to say that the occurrence of a certain number of "frauds" in the population would result, to any required probability, in at least one occurrence in the sample.

9. Estimation. Perhaps the most common objective of sampling is to provide estimates of certain properties of the population. Typically we may be interested in the mean value of a population of numerical values, or in its variance. Alternatively, we may wish to compare the properties of one population with those of another, and to do so using a sample from each population. As the accountant for a group of supermarkets, for instance, we may wish—using samples of customers —to estimate the average expenditure of a customer in the supermarket, the variance in individual expenditure, and the difference in average customer expenditure between two supermarkets.

It is with estimation that we shall be mainly concerned in the following sections.

ESTIMATION

Throughout the subsequent work it is assumed that all samples are drawn by the process of simple random sampling. We shall also assume for the moment that the sample forms a negligible proportion of the population. This is similar to assuming the population to be infinite.

10. Estimation of mean. Suppose that a sample is drawn from a population with the objective of estimating the population mean value. A natural way of estimating this parameter would be to calculate the mean value of the sample, *i.e.* the *sample mean*. Introducing notation which we shall use frequently, the process of estimation is summarised by: Population mean, μ, is estimated by the sample mean $\bar{x}$, where

$$\bar{x} = (x_1 + \ldots + x_n)/n \qquad (i)$$

and

$x_1, \ldots, x_n$ are the n sample values.

To distinguish between (i) and the actual numerical result which is obtained when the sample values are inserted in (i), the former, $\bar{x}$, is referred to as an *estimator* of μ, and the latter an *estimate* of μ.

The estimate which is calculated from the sample will typically not be exactly the value of the population parameter, μ. Even if it were, we would have no way of knowing, without recourse to examining the whole population and thereby defeating the object of sampling. Accordingly our estimate will be subject to sampling error, and ideally we should like to know how large this is.

Consider the state of affairs *before* sampling. We know that there are a large number of ways in which our sample could be drawn, and that each of the possible resulting samples will have a mean. Consequently we may think conceptually of our set of possible samples in terms of the mean values they generate. This is illustrated in Fig. 48 in general, and with reference to a particular simple case.

The remarkable fact is that if we were to present all the sample means in the form of a relative frequency histogram, the result would be approximately a normal distribution with mean μ. To some extent this result may not be totally unexpected, since one can appreciate that some samples will have a mean less than μ, some greater, but "on average" that the sample means will group in a symmetrical way about μ. The closeness of the approximation to normality depends on the size of the samples, n, and on the shape of the population distribution. (The latter, of course, we usually do not know.) In fact, the larger the value of n, the better the approximation to normality for a given population distribution. If the population is itself normal, then the distribution of the sample means is *exactly* normal (irrespective of whether n is large or not). Usually the population will not have a normal distribution (although it may often be fairly near) and it is then difficult to give hard and fast rules as to how large n must be for the approximation to be satisfactory. As a general guide, however, for

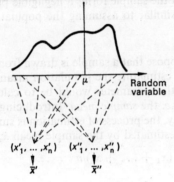

(a) Different sample means resulting from drawing of different samples from population.

Population: 20, 15, 18, 19, 17, 12, 23.
$\mu = 17 \cdot 71$
$n = 2$
Possible samples:

mean	mean	mean
20, 15 → 17·5	15, 19 → 17·0	18, 23 → 20·5
20, 18 → 19·0	15, 17 → 16·0	19, 17 → 18·0
20, 19 → 19·5	15, 12 → 13·5	19, 12 → 15·5
20, 17 → 18·5	15, 23 → 19·0	19, 23 → 21·0
20, 12 → 16·0	18, 19 → 18·5	17, 12 → 14·5
20, 23 → 21·5	18, 17 → 17·5	17, 23 → 20·0
15, 18 → 16·5	18, 12 → 15·0	12, 23 → 17·5

(b) Sample means which might result in drawing a sample of size two from given population.

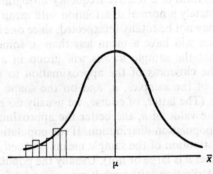

(c) Limiting curve of relative frequency histogram of sample means.

Fig. 48.—Variation in sample mean.

the reasonably well-behaved population distributions which we are likely to meet, for $n > 20$ the normal approximation to the *sampling distribution* of the sample mean is quite satisfactory.

Although we have explained that the sample mean has a normal distribution centred on μ we have not referred to its variance. Both mean and variance can be found from:

$$\bar{x} = (x_1 + \ldots + x_n)/n$$

since $x_1, \ldots, x_n$ are independent random variables with mean μ, and variance equal to the population variance. Thus,

$$E(\bar{x}) = (\mu + \ldots + \mu)/n = \mu$$
$$\text{var}\,(\bar{x}) = (\text{var}\,x_1 + \ldots + \text{var}\,x_n)/n^2$$
$$= \sigma_{\bar{x}}^2, \text{ say}$$
$$= \sigma^2/n.$$

The second result is interesting in that the larger the value of n (*i.e.*, sample size), the smaller is the variance of the sample mean (*see* Fig. 49). This is to be expected, since the larger the sample, the more accurate is our estimate likely to be.

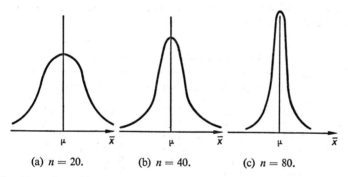

(a) $n = 20$.   (b) $n = 40$.   (c) $n = 80$.

FIG. 49.—Effect of increased sample size, n, on distribution of sample mean.

The ideas that we have been discussing are embodied in a very important statistical result known as the *Central Limit Theorem*. This states that, as n tends to infinity (*i.e.*, becomes progressively larger), the probability distribution of $(\bar{x} - \mu)/\sigma_{\bar{x}}$ approaches the standard normal.

Knowledge of the sampling distribution of the mean enables us to make precise statements about the degree of confidence we place in our estimate. Thus we know that the probability of a standard normal variable lying between $\pm 1\cdot 96$ is $0\cdot 95$, *i.e.*

$$-1\cdot 96 < \frac{\bar{x} - \mu}{\sigma/\sqrt{n}} < 1\cdot 96$$

or
$$-1{\cdot}96\sigma/\sqrt{n} < \bar{x} - \mu < +1{\cdot}96\sigma/\sqrt{n}$$

or
$$\bar{x} - 1{\cdot}96\sigma/\sqrt{n} < \mu < \bar{x} + 1{\cdot}96\sigma/\sqrt{n}$$

with a probability of 95%. This is called a *95% confidence interval for* μ, since it gives limits within which μ lies with a probability of 95%.

To specify this interval in practice we need values for $\bar{x}$ (known), for n (known), and for σ (unknown). Fortunately, not too much error is incurred by using the sample variance as a substitute for the population variance σ^2. Thus σ^2 is estimated by

$$s^2 = \frac{\Sigma x_i^2 - (\Sigma x_i)^2/n}{(n-1)}.$$

Notice that a divisor of $(n-1)$ is used here, since it can be shown that a divisor of n results in a general underestimation of σ^2.

Worked Example

A sample of 50 invoices is selected from a large population. Their values are given in Table 25. Find (a) 95%, (b) 90%, (c) 99% confidence intervals for the population mean.

Table 25. Random sample of invoices

Values of 50 sampled invoices, x_i ($i = 1, ..., 50$)				
2·53	1·87	2·27	8·05	3·38
1·47	0·05	4·11	5·11	2·43
0·52	3·48	4·86	8·00	9·04
0·25	1·49	2·29	3·39	9·37
3·16	0·84	8·12	6·02	3·84
1·98	4·51	2·20	7·51	4·59
2·10	1·75	8·07	9·07	2·04
0·37	7·38	5·27	6·06	6·78
3·01	8·05	4·95	2·93	7·39
2·41	9·05	8·06	5·59	0·97

Answer

Sample mean
$$= \Sigma x_i/n = 218{\cdot}03/50 = 4{\cdot}361 = \bar{x}$$
$$\Sigma x_i^2 = 1335{\cdot}58.$$

Variance of sample
$$= \frac{\Sigma x_i^2 - n\bar{x}^2}{(n-1)} = 7{\cdot}70.$$

Variance of sample mean

$$= 7\cdot70/50$$
$$= 0\cdot154.$$

(a) 95% confidence interval for μ·

$$4\cdot361 - 1\cdot96 \ \sqrt{(0\cdot154)} < \mu < 4\cdot361 + 1\cdot96 \ \sqrt{(0\cdot154)}$$

i.e.,

$$3\cdot59 < \mu < 5\cdot13.$$

(b) 90% confidence interval for μ:

$$4\cdot361 - 1\cdot645 \ \sqrt{(0\cdot154)} < \mu < 4\cdot361 + 1\cdot645 \ \sqrt{(0\cdot154)}$$

i.e.,

$$3\cdot716 < \mu < 5\cdot006.$$

(We use the value of $1\cdot64$ since with 90% probability a standard normal variable lies between $\pm1\cdot64$.)

(c) 99% confidence interval for μ:

$$4\cdot361 - 2\cdot58 \ \sqrt{(0\cdot154)} < \mu < 4\cdot361 + 2\cdot58 \ \sqrt{(0\cdot154)}$$

i.e.,

$$3\cdot348 < \mu < 5\cdot372.$$

These results are illustrated in Fig. 50. Notice that as the degree of confidence required is raised, the length of the confidence interval is increased. This is to be expected, since we are wanting to be increasingly sure of having an interval which includes μ.

Looking at the construction of confidence intervals from a slightly different viewpoint, we may say that of all the possible samples which could be drawn, 95% will have a mean value $\bar{x}$ such that

$$-1\cdot96 < \frac{\bar{x} - \mu}{\sigma/\sqrt{n}} < 1\cdot96. \qquad (i)$$

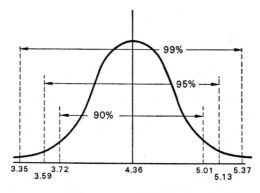

FIG. 50.—Confidence intervals for μ.

This being the case, the probability that a particular random sample will have a mean value satisfying (i) is 95%, and therefore

$$\bar{x} - 1.96\sigma/\sqrt{n} < \mu < \bar{x} + 1.96\sigma/\sqrt{n}$$

with a probability of 95% (or 0·95).

Incidentally, we have talked exclusively of confidence intervals which are symmetric about the true population value, μ. Although asymmetric confidence intervals are rarely used in practice, in theory there would be nothing incorrect in determining a 95% confidence interval, say, in which 90% of the probability was above $\bar{x}$, and 5% below.

We have assumed that the sample size n is always known. However, n is under our control and we have to decide before sampling what is an appropriate value to take. In fact n depends on (a) the precision with which our estimate is required (i.e., the narrowness of the confidence interval) and (b) the probability with which this precision is to be obtained.

Suppose, for example, in assessing the spending power of a community, that it is required to find the average weekly per capita expenditure on "luxury" goods. The estimate is to be within ± 10p of the true value, with a probability of 97%. We know that

$$\frac{\bar{x} - \mu}{\sigma/\sqrt{n}} \sim N(0, 1)$$

and therefore that

$$-2.17\sigma/\sqrt{n} < (\bar{x} - \mu) < 2.17\sigma/\sqrt{n}$$

with a probability of 97%. We require then, that

$$2.17\sigma/\sqrt{n} \leqslant 10.$$

If σ is known from past experience to be £1·00, it follows that

$$n \geqslant \{(2.17 \times 100)/10\}^2 = 470.9.$$

A sample size of at least 471 is required for the desired precision and confidence.

If σ is unknown, a pilot sample of size 25, say, could be taken to obtain an estimate. Alternatively, as a simple rule of thumb, the likely range of the population—say £2–£10—can be used to estimate the standard deviation, since it usually extends to roughly 3σ on either side of the mean. Thus £8 $\doteq$ 6σ, or $\sigma \doteq$ £1·33.

Worked Example

Unicorn Ltd. is considering buying a large quantity of components from Universal Electronics Ltd. The sales manager of Universal Electronics is requested to provide Unicorn with a figure for the mean life of the components. He considers it worth £4000 to obtain an estimate that has 19 chances in 20 of being within 0·05 of the correct value. The cost of setting up equipment to test the components is £2500, and the cost of testing a component to destruction is £2·50. It is known from past records that the standard deviation of the life of components is 0·80.

Will Universal Electronic's sales manager be able to obtain the required estimate for £4000 or less? What is the minimum cost of obtaining the necessary estimate?

Answer

The sum of money available for the investigation is £4000. The size of sample, n, which may be tested within this budget is given by:

$$n = (4000 - 2500)/2·50 = 600.$$

Suppose $\bar{x}$ is the mean of a sample of 600, μ is the population mean, and σ is the population standard deviation. We seek a 95% confidence interval ("19 chances in 20" . . .).

Thus,
$$\bar{x} - \frac{1·96\sigma}{\sqrt{n}} < \mu < \bar{x} + \frac{1·96\sigma}{\sqrt{n}}$$

or
$$-\frac{1·96\sigma}{\sqrt{n}} < \bar{x} - \mu < \frac{1·96\sigma}{\sqrt{n}}$$

i.e.,
$$-\frac{1·96 \times 0·80}{\sqrt{600}} < \bar{x} - \mu < \frac{1·96 \times 0·80}{\sqrt{600}}$$

i.e.,
$$-0·064 < \bar{x} - \mu < 0·064,$$

with a probability of 95%.

Our conclusion then, is that with a sample size of 600 (*i.e.*, a cost of £4000) the difference between sample mean and population mean, with a probability of 95%, can be as much as 0·064. This is not adequate. The only variable under his control is the sample size n. He requires that

$$-0·05 < \bar{x} - \mu < 0·05$$

with a probability of 0·95.

Therefore,
$$\frac{1·96 \times 0·80}{\sqrt{n}} \leqslant 0·05$$

i.e.,
$$n \geqslant \{(1·96 \times 0·80)/0·05\}^2$$
$$\geqslant 983·4.$$

For the required precision, a sample of at least 984 is needed. The cost of investigating such a sample is:

$$£2500 + (984 \times 2·50) = £4960.$$

Exercises on Section 10

1. In evaluating various plans and funding methods, a corporation must determine the mean age of its work force. Since the company has several thousand employees, a sample must be taken. Estimate how large the sample must be in order that the estimate of the mean may be incorrect by not more than 2 years with 90% confidence, if the standard deviation of ages is taken to be 10 years.

2. Packages to be delivered from a warehouse have a mean weight of 450 lb, and a standard deviation of 75 lb. What is the probability that 37 packages taken at random and loaded on to a truck will exceed the capacity of the truck, which is 12,300 lb?

3. The average time in days required to process orders received by a metal fabrication company is to be estimated. A sample of 50 orders is selected randomly from a total of 375 orders processed over a period of six months. The sample mean is 5·4 days, and the sample S.D. is 1·4 days. Compute the 95% interval estimate of the average processing time.

4. The following information is given about the present accounts receivable of a certain corporation: (1) 10% are accounts which have been classified as dubious as regards rapidity of collection; (2) The sizes of the accounts are normally distributed with mean £10,000 and S.D. £2000.

(*a*) If five accounts are randomly selected, what is the probability of obtaining exactly one dubious account?

(*b*) If one account is selected, what is the probability that it lies between £9000 and £12,000 in size?

(*c*) If a random sample of 400 accounts is selected, what is the probability that the sample mean lies between £10,000 and £10,200 in size?

(*d*) If one account is selected at random, the probability is 0·15 that its size will exceed a certain amount. What is this amount?

5. Dermist, the owner of Dermist's Taxi fleet, is to place an order for Shedtread Tyres. He requires an estimate of the life of the tyres, and over the past has found that the 76 tyres of this make that he has used have had an average life of 21,000 miles, with a standard deviation of 1500 miles. Construct a 98% confidence interval for the average life of a Shedtread tyre.

6. In Question 5 above, what sample size would have been necessary to estimate the average lifetime to within 200 miles of the true value, with a probability of 0·98?

7. A manufacturer has sales offices in four major cities. Each office has 25 salesmen. The weekly sales of any salesman are normally distributed with a mean of £1200 and a S.D. of £200. Within what range about the mean is the probability 0·997 that

(*a*) a given salesman's weekly sales will occur?

(*b*) the average weekly sales per salesman in a particular office will occur?

(*c*) the average weekly sales per salesman of the company will occur?

8. The average number of photocopies made per working day in a certain office is 356 with a S.D. of 55. It costs the firm 3p per copy. During

a working period of 121 days, what is the probability that the average cost per day is more than £11·10?

9. What is the probability of drawing a simple random sample of size 100 which has a mean of 30 or more from a population with a mean of 28? The population variance is 81. Do you have to assume the population is normal?

10. A sample of 45 items from the first day's output of a new machine is taken, and the processing time of each item on the machine noted. The sample mean is 1·020 minutes with standard deviation of 0·030 minutes. Find a 97% confidence interval for the population mean.

11. A company manufacturing radios selected a random sample of 50 radios in order to estimate the mean number of defects per radio, the results being as follows:

No. of defects	No. of radios
0	24
1	13
2	6
3	4
4	2
5	1

Estimate the mean number of defects per radio with a 95% confidence interval.

11. Estimation of the mean: small samples.

In the previous section we claimed that, when n was large, the statistic $(\bar{x} - \mu)/(\sigma/\sqrt{n})$ had approximately a standard normal distribution. Although exact normality is achieved only when the population itself is normal (and is then true for all values of n, small or large), when $n \geqslant 20$ the approximation is very adequate for most purposes. Indeed, we went further, and stated that the standard normal approximation was valid for the statistic $(\bar{x} - \mu)/(s/\sqrt{n})$ which is used when the population variance σ^2 is unknown.

Unfortunately this happy state of affairs does not exist when n is small (i.e., <20). In this case, if the population distribution is far from normal, and not known precisely, there is little one can do. However, if the population can be sensibly assumed to have a distribution which is roughly normal and not markedly skewed we can modify our previous argument.

For small samples it is not sufficient to assume that the sample variance, s^2, is a reliable estimator of σ^2. We must take into account the fact that s is based on sample information, and therefore has a sampling distribution of its own. The effect of this was first studied by W. W. Gosset, writing under the pseudonym of "Student." He concluded that the statistic has (when the population is near normal) a distribution which he called a *t-distribution*.

There is in fact a family of such distributions all symmetrical and with means zero, but with a shape dependent on the parameter $(n - 1)$. The *t*-distribution is compared with the standard normal in Fig. 51.

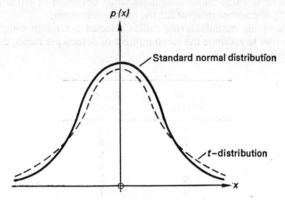

FIG. 51.—Comparison of standard normal and Student's *t*-distribution.

The parameter $(n - 1)$ is called the number of *degrees of freedom* of the particular *t*-distribution. The concept of degrees of freedom is a rather abstract one, occurring not only in statistical work but in many other branches of applied mathematics. In the present context it is perhaps most easily interpreted by noting that, for a fixed value of $\bar{x}$, there are only $(n - 1)$ variables which may alter independently in calculating s^2. Thus, given $\bar{x}$, $x_1 \ldots, x_{n-1}$ can assume any values, but once those values are fixed, x_n—and s^2—is automatically determined. In other words, in the calculation of s^2 the variables $x_1, \ldots, x_n$ which are otherwise free to vary independently, are placed under one constraint, *i.e.* $(x_1 + \ldots + x_n)/n = \bar{x}$.

The effect, then, on our determination of confidence intervals is that we must use the *t*-distribution tables with the appropriate number of degrees of freedom rather than the normal tables.

Worked Examples

1. In the first worked example of Section 10 above, suppose the sample of invoices had been of size 10, and in fact had been the last column of Table 25, *i.e.*

3·38, 2·43, 9·04, 9·37, 3·84, 4·59, 2·04, 6·78, 7·39, 0·97.

Find a 95% confidence interval for μ.

Answer

$$\text{Sample mean} \quad = 4\cdot983$$

$$\text{Sample variance} = \frac{\Sigma x_i{}^2 - (\Sigma x_i)^2/n}{n-1}$$

$$= \frac{328\cdot35 - 248\cdot30}{9}$$

$$= \frac{80\cdot05}{9}$$

$$= 8\cdot89.$$

Thus,

$$\left| \frac{4\cdot983 - \mu}{\sqrt{\left(\frac{8\cdot89}{10}\right)}} \right| < t_9 \, (95\%).$$

Consulting the *t*-distribution tables for 9 degrees of freedom, we find that 95% of the area under the curve lies between $\pm 2\cdot26$ (*see* Appendix IV).
   Thus,

$$4\cdot983 - 2\cdot26 \, \sqrt{(0\cdot889)} < \mu < 4\cdot983 + 2\cdot26 \, \sqrt{(0\cdot889)}$$
$$2\cdot852 < \mu < 7\cdot114.$$

Note that the 95% confidence interval is much wider than that based on a sample size of 50. The obvious reason for this is our use of the *t*-value of $2\cdot26$, rather than the normal value of $1\cdot96$. More practically this difference arises because our confidence interval depends on an estimate of the population variance, which is less reliable in the case of a small sample. This in turn means that we are less sure of our estimate of μ.

2. Find a 98% confidence interval estimate of the mean income of a random sample of 17, given that the sample mean $\bar{x} = £1850$, and $s = £250$.

Answer

Since $n = 17$, we consult the tables for the *t*-distribution with 16 degrees of freedom. Thus $t_{16} \, (98\%) = 2\cdot58$,

and

$$\bar{x} - \frac{2\cdot58 \times 250}{\sqrt{17}} < \mu < \bar{x} + \frac{2\cdot58 \times 250}{\sqrt{17}}$$

i.e.,

$$1850 - 156\cdot44 < \mu < 1850 + 156\cdot44$$
$$1693\cdot56 < \mu < 2006\cdot44.$$

Exercises on Section 11

12. A working design is under consideration for adoption in a plant. A time and motion study shows that 11 workers have a mean assembly time of 294 seconds, with a standard deviation of 12 seconds. Provide a 90% confidence interval for the mean assembly time.

13. The time required by an operator to perform a particular job is normally distributed. Five observations were made on a particular operator as follows:

$$17 \cdot 9 \text{ sec., } 17 \cdot 3 \text{ sec., } 18 \cdot 4 \text{ sec., } 18 \cdot 3 \text{ sec., } 17 \cdot 7 \text{ sec.}$$

Estimate the mean time to perform the job, with a 99% confidence interval.

14. A portfolio manager is running a 12-share portfolio. The average value of the rates of return on the 12 shares, and their standard deviation, are given below:

$$\bar{x} = 13 \cdot 7\%.$$
$$s = 2 \cdot 9\%.$$
$$n = 12.$$

Assuming historic returns can be used to predict future returns, find a 96% confidence interval for the portfolio's future return.

15. A manufacturing company, Slipup Ltd., insures with the Accident Prone Insurance Co. for claims for industrial injury by Slipup employees against their company. Accident's actuaries want to quote a realistic premium to Slipup, and examine the 15 claims of this type that they have had to settle over the last year. The settlements (£) were respectively for:

2100	2914	3728	485	197
4708	2571	1976	726	384
2819	1493	1110	548	897

Find a 95% confidence interval for the average claim.

12. Estimation of the difference between means: large samples.

We have been concerned so far with estimation on the basis of a single sample from one population. However, it is very useful to be able to compare samples taken from different populations. Suppose, for instance, that we have samples of sizes n_1 and n_2 respectively drawn from populations with mean values μ_1 and μ_2, and variances σ_1^2 and σ_2^2. Denoting by $x_{11}, x_{21}, \ldots, x_{n_1 1}$ the first sample, and by $x_{12}, x_{22}, \ldots, x_{n_2 2}$ the second, we have the following situation:

Population I
Mean μ_1
Variance σ_1^2

Population II
Mean μ_2
Variance σ_2^2

Sample I
Size n_1
Mean $\bar{x}_1 = \dfrac{x_{11} + \ldots + x_{n_1 1}}{n_1}$
Variance $s_1^2 = \dfrac{\Sigma x_{i1}^2 - n_1 \bar{x}_1^2}{n_1 - 1}$

Sample II
Size n_2
Mean $\bar{x}_2 = \dfrac{x_{12} + \ldots + x_{n_2 2}}{n_2}$
Variance $s_2^2 = \dfrac{\Sigma x_{i2}^2 - n_2 \bar{x}_2^2}{n_2 - 1}$

In order to estimate the difference in population means, $\mu_1 - \mu_2$, we would naturally use the estimator $\bar{x}_1 - \bar{x}_2$.

Now, if n_1, n_2 are both large,

$$\bar{x}_1 \sim N(\mu_1, \sigma_1^2)$$
$$\bar{x}_2 \sim N(\mu_2, \sigma_2^2).$$

Thus, $\bar{x}_1 - \bar{x}_2$ will have a mean value $\mu_1 - \mu_2$, and variance

$$\left(\frac{\sigma_1^2}{n_1} + \frac{\sigma_2^2}{n_2} \right),$$

on the assumption that the samples are independent (*see* Chapter 4, Section 14). Indeed, it can be shown that the sum or difference of two normally distributed random variables is itself normal, so

$$(\bar{x}_1 - \bar{x}_2) \sim N\left(\mu_1 - \mu_2, \frac{\sigma_1^2}{n_1} + \frac{\sigma_2^2}{n_2} \right)$$

i.e.,

$$\frac{(\bar{x}_1 - \bar{x}_2) - (\mu_1 - \mu_2)}{\sqrt{\left\{ \frac{\sigma_1^2}{n_1} + \frac{\sigma_2^2}{n_2} \right\}}} \sim N(0, 1).$$

We may now use our previous argument (Section 10 above) to find confidence intervals for $(\mu_1 - \mu_2)$. A 95% interval, for example, is given by:

$$(\bar{x}_1 - \bar{x}_2) - 1 \cdot 96 \sqrt{\left(\frac{\sigma_1^2}{n_1} + \frac{\sigma_2^2}{n_2} \right)} < \mu_1 - \mu_2 < (\bar{x}_1 - \bar{x}_2)$$
$$+ 1 \cdot 96 \sqrt{\left(\frac{\sigma_1^2}{n_1} + \frac{\sigma_2^2}{n_2} \right)}.$$

Typically σ_1^2, σ_2^2 will be not known, and estimates s_1^2, s_2^2 provided by the sample variances will have to be used. The 95% confidence interval will then be:

$$(\bar{x}_1 - \bar{x}_2) - 1 \cdot 96 \sqrt{\left(\frac{s_1^2}{n_1} + \frac{s_2^2}{n_2} \right)} < \mu_1 - \mu_2 < (\bar{x}_1 - \bar{x}_2)$$
$$+ 1 \cdot 96 \sqrt{\left(\frac{s_1^2}{n_1} + \frac{s_2^2}{n_2} \right)}.$$

Worked Example

A footwear manufacturer is contemplating the purchase of an expensive additional machine. Two rival machines, the Bootleg and the Slipshod are on the market, and both suit his manufacturing requirements. He installs one machine of each type on a test basis, and runs them for 23 days and 29 days respectively. Output figures are:

	Bootleg	Slipshod
Daily average	$\bar{x}_1 = 140$	$\bar{x}_2 = 120$
Standard deviation of daily output	$s_1 = 20$	$s_2 = 10$
Sample size	$n_1 = 23$	$n_2 = 29$

Find a 98% confidence interval for the difference in average daily output between the two machines.

Answer

For a 98% confidence interval the tabulated value of the standard normal variable is 2·33. Thus, the required interval estimate is:

$$(140 - 120) - 2\cdot33 \sqrt{\left\{\frac{400}{23} + \frac{100}{29}\right\}} < \mu_1 - \mu_2 < (140 - 120)$$
$$+ 2\cdot33 \sqrt{\left\{\frac{400}{23} + \frac{100}{29}\right\}}$$

i.e.,

$$9\cdot36 < \mu_1 - \mu_2 < 30\cdot64.$$

13. Estimation of difference between means: small samples. Since we ran into a little difficulty when estimating the population mean using a small sample, it should not come as any surprise to learn that problems arise in comparing two means when the samples are small. In fact in general there is no simple way of finding a confidence interval for $(\mu_1 - \mu_2)$ in the small sample case, since variables with t-distributions do not combine in the convenient way enjoyed by normally distributed variables.

However, we can make some progress if two populations have the same variance.

Using the notation of Section 10 above we firstly obtain an estimator, s^2, of the common variance, σ^2, in a natural way,

viz.

$$s^2 = \frac{\Sigma(x_{i1} - \bar{x}_1)^2 + \Sigma(x_{i2} - \bar{x}_2)^2}{n_1 + n_2 - 2} \qquad (i)$$

i.e.,

$$s^2 = \frac{(n_1 - 1)s_1^2 + (n_2 - 1)s_2^2}{n_1 + n_2 - 2} \qquad (ii)$$

This estimator is known as the *pooled variance*. The numerator in (i) is the total squared deviation from the respective sample means. The explanation for the denominator used rests on the concept of *degrees of freedom*. The $(n_1 + n_2)$ variables $x_{11}, \ldots, x_{n_11}, x_{12}, \ldots, x_{n_22}$ upon which s^2 depends are subject to the two constraints:

$$(x_{11} + \ldots + x_{n_11})/n_1 = \bar{x}_1$$

and

$$(x_{12} + \ldots + x_{n22})/n_2 = \bar{x}_2.$$

These reduce the number of degrees of freedom from $(n_1 + n_2)$ to $(n_1 + n_2 - 2)$.

In form *(ii)*, s^2 is seen to be a weighted average of the two independent estimators s_1^2, s_2^2 in which greater weight is attached to the estimator arising from the larger sample.

The procedure for constructing confidence interval estimates of $(\mu_1 - \mu_2)$ now follows the familiar pattern:

$$\bar{x}_1 \sim N(\mu_1, s_1^2/n_1)$$
$$\bar{x}_2 \sim N(\mu_2, s_2^2/n_2).$$

If the samples are independent we may then deduce that:

$$(\bar{x}_1 - \bar{x}_2) \sim N\left(\mu_1 - \mu_2, s^2\left(\frac{1}{n_1} + \frac{1}{n_2}\right)\right)$$

i.e.,

$$\frac{(\bar{x}_1 - \bar{x}_2) - (\mu_1 - \mu_2)}{\sqrt{\left\{s^2\left(\frac{1}{n_1} + \frac{1}{n_2}\right)\right\}}} \sim t_{n_1+n_2-2}.$$

Worked Example

The average weekly power costs of two factories are to be compared. Independent samples of sizes 12 and 9 are taken at random from the records kept over the last year. The figures are given below:

Factory I		Factory II	
210	209	187	182
229	215	165	185
190	225	184	204
223	198	200	167
192	201	190	
207	212		

Find a 95% confidence interval for $(\mu_1 - \mu_2)$.

Answer

	Factory I	Factory II
Mean	$\bar{x}_1 = 2511/12$	$\bar{x}_2 = 1664/9$
	$= 209{\cdot}25$	$= 184{\cdot}89$
Variance	$s_1^2 = 157{\cdot}91$	$s_2^2 = 168{\cdot}15$
Sample size	$n_1 = 12$	$n_2 = 9$

$$\text{Pooled variance, } s^2 = \frac{1737 + 1345}{12 + 9 - 2} = 162{\cdot}21$$

For t_{19}, the 95% points are $\pm 2{\cdot}093$.

The required 95% confidence interval is given by:

$$(209{\cdot}25 - 184{\cdot}89) - 2{\cdot}093 \sqrt{\left\{162{\cdot}21\left(\frac{1}{12} + \frac{1}{9}\right)\right\}} < \mu_1 - \mu_2$$

$$< (209{\cdot}25 - 184{\cdot}89) + 2{\cdot}093 \sqrt{\left\{162{\cdot}21\left(\frac{1}{12} + \frac{1}{9}\right)\right\}}$$

i.e.,

$$12{\cdot}61 < \mu_1 - \mu_2 < 36{\cdot}11.$$

Exercises on Section 13

16. The annual salary of executives in a certain city averages £4000 and has a standard deviation of £300. In the same city, the annual salary of physicians averages £4500 and has a standard deviation of £450. A random sample of 100 is taken from each population. What is the probability that the sample means differ (*a*) by less than £1200? (*b*) by more than £1500?

17. A second working design is under consideration (*see* Question 12, Exercises on Section 11 above) and the 15 workers using this design have an assembly time with a mean of 340 seconds, and a standard deviation of 16 seconds. Using a pooled variance estimate, find a 98% confidence interval estimate for the difference in mean times.

18. The street lighting authorities bought 100 bulbs of type *A*, and another 100 of type *B*. On testing these bulbs, they found that $\bar{x}_A = 1300$ hours, $s_A = 90$ hours, $\bar{x}_B = 1250$ hours, and $s_B = 100$ hours. What is the probability that the difference between the two corresponding populations is greater than 40 hours?

19. Two different makes of T.V. picture tubes *A* and *B* possess the following parameters: $\mu_A = 1400$ hours, $\sigma_A{}^2 = 40{,}000$ hours; $\mu_B = 1200$ hours; $\sigma_B{}^2 = 10{,}000$ hours. A random sample of 125 tubes is drawn from each make. Determine the probability that (*a*) make *A* will have a mean life at least 160 hours longer than *B*: (*b*) make *A* will have a mean life at least 250 hours longer than *B*.

20. The Seafood Co. Ltd. contemplates investing in a new trawler to expand its fishing operations. Investment may be made in one of two trawlers, *A* and *B*. In the case of *A*, £7000 will be invested and the net cash inflows generated from years 1–5 will each be normally distributed with mean £2000, and standard deviation £400. In the case of *B*, £5000 will be invested and the net cash inflows from years 1–5 will each be normally distributed with mean £1500 and standard deviation £300.

Calculate the expected values and variances of the NPV of each investment, assuming that the cash inflows are independent of each other and that the discount rate is 12%.

(*a*) Which investment would you prefer, and why?

(*b*) Provided with the further information that the two standard deviations quoted above have each been based on values of 5 cash inflows for past projects very similar to *A* and *B* respectively give a 90% confidence interval for the difference between the NPVs of the cash inflows in year 1.

21. Random samples of male and female customers were selected on leaving a supermarket, and the amounts each individual had spent were determined. The results were:

Male	Female
$\bar{x}_1 = £1 \cdot 28$	$\bar{x}_2 = £2 \cdot 02$
$s_1 = £0 \cdot 31$	$s_2 = £0 \cdot 43$
$n_1 = 10$	$n_2 = 12$

Calculate the pooled variance, and find a 97% confidence interval estimate for the difference between male and female average expenditure.

14. Estimation of proportion: large sample.

It is often useful to know what proportion of a population has a particular attribute. Thus before an election involving two candidates, a sample of the electorate may be asked if they will vote for candidate A or not. The object of the exercise is to estimate the proportion of the total electorate who will vote for candidate A. As a second illustration, suppose that an accountant wishes to determine the proportion of customers who are incorrectly invoiced. By taking a sample of invoices before despatch he can estimate the proportion of the invoices which contain errors.

In general, suppose we have taken a sample of size n, and that the number of items in the sample with the particular property in which we are interested is r. We take r/n as an estimate of the population proportion, p.

Remembering our earlier work on the binomial distribution, the probability of obtaining exactly r qualifying items in a sample size n, when the probability of any individual item qualifying is p, can be calculated from:

$$\frac{n!}{r!\,(n-r)!}\,p^r(1-p)^{n-r}.$$

Let R be the random variable "number of items with required property in sample of size n." When n is large, we recall that the binomial distribution can be approximated to by the normal distribution with mean $\mu = np$, and variance $\sigma^2 = np(1-p)$,

i.e.,

$$R \sim N(np, np(1-p))$$

and

$$\frac{R}{n} \sim N\left(p, \frac{p(1-p)}{n}\right).$$

Now R/n is the sample proportion with the required property, and we conclude that—as long as p is not too near 0 or 1—

$$\left(\frac{R}{n} - p\right)\bigg/\sqrt{\left\{\frac{p(1-p)}{n}\right\}} \sim N(0, 1).$$

Following the now familiar argument we may construct, say, a 98 % confidence interval for p as:

$$\frac{r}{n} - 2\cdot33\sqrt{\left\{\frac{p(1-p)}{n}\right\}} < p < \frac{r}{n} + 2\cdot33\sqrt{\left\{\frac{p(1-p)}{n}\right\}}.$$

Unfortunately, of course, the variance $p(1-p)/n$ is not known, since p is the very quantity that we are hoping to estimate. Two ways of resolving this problem are available.

Firstly we may use r/n as a substitute for p without incurring too much error (cf. our substitution of s^2 for σ^2 when estimating means). Thus (i) is modified to become:

$$\frac{r}{n} - 2\cdot33\sqrt{\left\{\frac{r}{n}\left(1-\frac{r}{n}\right)\Big/n\right\}} < p < \frac{r}{n} + 2\cdot33\sqrt{\left\{\frac{r}{n}\left(1-\frac{r}{n}\right)\Big/n\right\}}.$$

Alternatively, whatever the value of p (which, by definition lies between 0 and 1), the product $p(1-p)$ can never be greater than $\frac{1}{4}$:
If

$$p = \tfrac{1}{4}, p(1-p) = \tfrac{3}{16} < \tfrac{1}{4},$$
$$p = \tfrac{1}{3}, p(1-p) = \tfrac{2}{9} < \tfrac{1}{4}, \text{ etc.}$$

Consequently, the variance $p(1-p)/n$ is never greater than $\frac{1}{4}n$, and the 98 % confidence interval is certainly not wider than

$$\frac{r}{n} - 2\cdot33\sqrt{\left(\frac{1}{4n}\right)} < p < \frac{r}{n} + 2\cdot33\sqrt{\left(\frac{1}{4n}\right)}.$$

This 98 % confidence interval is "safe," in the sense that if p is not exactly equal to $\frac{1}{2}$, it could be narrowed.

Worked Example

Before deciding whether to stock a new line in ladies night attire a boutique is interested in knowing what proportion of the items are returned by the customers as faulty. The store's manager knows that faulty goods are costly in terms of loss of customer goodwill, and in the work caused by returning the goods to the manufacturers. Accordingly a batch of 150 is purchased for a trial period. It is found that 22 of these are returned. Find a 96 % confidence interval for the proportion of defectives.

Answer

$r = 22, n = 150$.

From normal tables, 96 % of the area lies between $\pm 2\cdot05$.

Thus:

$$\frac{22}{150} - 2\cdot05 \sqrt{\left\{ \frac{\frac{22}{150}\left(1 - \frac{22}{150}\right)}{150} \right\}} < p < \frac{22}{150}$$

$$+ 2\cdot05 \sqrt{\left\{ \frac{\frac{22}{150}\left(1 - \frac{22}{150}\right)}{150} \right\}}$$

i.e.,

$$0\cdot1467 - 0\cdot0592 < p < 0\cdot1467 + 0\cdot0592$$
$$0\cdot0875 < p < 0\cdot2059.$$

Alternatively a "safe" 96% confidence interval is given by:

$$0\cdot1467 - 2\cdot05 \sqrt{\left(\frac{1}{600}\right)} < p < 0\cdot1467 + 2\cdot05 \sqrt{\left(\frac{1}{600}\right)}$$

i.e.,

$$0\cdot0630 < p < 0\cdot2304.$$

When sampling for the purpose of estimating a proportion, the appropriate sample size for a required degree of precision and level of confidence can be predetermined (cf. Section 10 above). For accuracy within 0·03 of the true value with a probability of 90% we have

$$\left| \frac{r}{n} - p \right| < 1\cdot64 \sqrt{\left\{ \frac{p(1 - p)}{n} \right\}} \leqslant 0\cdot03$$

Conservatively,

$$1\cdot64 \sqrt{\left(\frac{1}{4n}\right)} \leqslant 0\cdot03$$

i.e.,

$$n \geqslant (1\cdot64/0\cdot03)^2/4 = 747\cdot1.$$

Thus a sample size of at least 748 is needed.

Exercises on Section 14

22. Similar products are produced by two manufacturers, A and B. A's output contains 7% defectives, and B's output, 5% defectives. If a random sample of 2000 is drawn from each manufacturer's product, what is the probability that the two samples will reveal a difference in proportional defectives of 0·01% or more?

23. As a ginger-ale producer entering a new territory, you need an estimate of the proportion of consumers that prefer to buy ginger-ale in cans. A consulting firm agrees to make a survey of ginger-ale buyers for £2000, plus £4 per interview. Assuming that they will use a random sample, and that the population proportion is equal to 0·50:

(a) How much will the survey cost if the error in estimating the proportion is to be no greater than 5 percentage points at the 90% confidence level?

(b) How much will the survey cost if the error is not to exceed 5 percentage points at the 98% confidence level?

24. If you wish to estimate the proportion of business executives who have graduate degrees and to have the estimate correct within 2% with a probability of 0·95 or better, how large a sample should be taken if you (*a*) are confident that the true proportion is less than 0·2 and (*b*) have no knowledge at all about the true proportion?

25. A sample of 100 customers' accounts is examined, and 10 are found to be "in the red." After circulation to all customers of a letter offering a discount for rapid payment of debts, a sample of 150 customers is examined and 12 are found to be "in the red." In the light of this information are you justified in thinking that your offer of discount has had any effect?

26. An insurance firm estimates that 15% of passengers purchase flight insurance in a particular airline terminal. An estimate of the true proportion of passengers who purchase flight insurance is to be computed so that the estimated proportion will differ from the population proportion by no more than 0·02 with a level of confidence of 95%. Compute the sample size required.

27. The general manager of a chain store is contemplating opening a new branch in the suburbs of Manchester. In the light of past information on sales, profits, and so on, he considers that if more than 35% of the customers who at present patronise the city centre store were to use the new branch, then the new branch would be viable. If, on the other hand, more than 40% of the present customers were to change their patronage, then the city centre store would cease to be viable. The total number of customers patronising the store is very large.

The general manager has £1000 available for market survey expenses, and approaches an organisation, Marketeer Ltd., to carry out a survey. Marketeer Ltd. normally charge a preparation fee of £200 and, in addition, a charge of £2 per customer interviewed. As the representative of Marketeer, what statistically-based argument would you present to the general manager to convince him that more money should be spent on the survey?

What increased figure would you quote him for the cost?

28. You are employed by a firm that must buy parts from a foreign distributor. It is known that, in the past, 20% of the parts secured from this distributor have been defective. The distributor claims that new techniques have reduced this figure below 20%. Your management places an order for 100,000 parts, and of a sample of 400, there are found to be 68 defectives.

What proportion of the time would the percentage defective in a random sample of this size differ due to chance from a universe value of 20% by as much (or more) than the difference you have observed?

15. Finite population correction.

In the preceding sections it has been assumed that the sample formed only a small fraction of the population. This is tantamount to supposing the population to be very large —effectively infinite—or to supposing that sampling is performed with replacement (*i.e.*, that a sampled item is replaced in the population immediately, so that it is a candidate for inclusion in the sample a second time, and the population is therefore not depleted by the sampling process).

The effect of having a finite population can be illustrated quite vividly. Suppose we have a population of 12 invoices, 6 of which are overdue for payment, and 6 of which have been paid. A sample of 6 is to be drawn to estimate the proportion of overdue invoices (true value 0·5). Suppose that the first invoice sampled is overdue. (This occurs with a probability of 0·50.) The probability that the second is overdue is now only $5/11 = 0·45$. This violates one of the conditions for the binomial distribution to be applicable, and therefore undermines the analysis of Section 14 above. If, of course, the first invoice sampled is replaced in the population for possible re-sampling the probability of selecting an overdue invoice would be maintained at 0·5.

The point is that our previous results have been built upon the assumption that the members of a sample are drawn independently. For an infinite population this is true, but for a finite population what has already been included in a sample may affect the probabilities of other members of the population being sampled.

Fortunately, it is not too difficult to correct for the presence of a finite population (or of sampling without replacement). The procedure is simply to incorporate the *finite population correction* factor, $\sqrt{\{(N-n)/(N-1)\}}$, into the expressions for the standard deviations of sample mean and sample proportion. Thus, with a population of size N,

$$\frac{\sigma}{\sqrt{n}} \rightarrow \sqrt{\left(\frac{N-n}{N-1}\right)} \cdot \frac{\sigma}{\sqrt{n}}$$

and

$$\sqrt{\left\{\frac{p(1-p)}{n}\right\}} \rightarrow \sqrt{\left(\frac{N-n}{N-1}\right)} \sqrt{\left\{\frac{p(1-p)}{n}\right\}}.$$

Since

$$\sqrt{\left(\frac{N-n}{N-1}\right)} = \sqrt{\left\{\frac{1-(n/N)}{1-(1/N)}\right\}}$$

and $1/N$ is small, the correction factor is approximately

$$\sqrt{\{1-(n/N)\}}.$$

The ratio is named the *sampling fraction*, since it gives the proportion of the population which is sampled.

Worked Example

In the first worked example of Section 10, suppose that the sample of 50 invoices had to be taken from a population of 250 invoices. Find the 95% confidence interval for μ.

Answer

The finite population correction is

$$\sqrt{\left(\frac{250-50}{250-1}\right)} = \sqrt{\left(\frac{200}{249}\right)} = 0.8962.$$

The confidence interval is therefore:

$$3.59 \times 0.8962 < \mu < 5.13 \times 0.8962$$
$$3.217 < \mu < 4.598.$$

CHAPTER 8

STATISTICAL HYPOTHESIS TESTING

INTRODUCTION

In the preceding chapter we were concerned with one aspect of inferential statistics—namely the making of statements about the values of population parameters, on the basis of information about part of the population. This chapter is devoted to a second form of inference in which sample information is used as a tool in decision-making.

A simple example of this involves the decision to accept or reject a consignment of goods on the basis of the number of defectives in a sample, or to accept or reject a batch of invoices on the basis of the number in a sample which contain errors. Alternatively, we may wish to test, on the basis of a sample, whether the average value of invoices is more, than, say £50, or whether two different populations have the same mean on the basis of information on two samples drawn from the populations. The methodology in testing properties of this nature is always the same, although the detailed analysis may vary with particular circumstances. The following example illustrates the principles of hypothesis testing.

TESTS CONCERNING POPULATION MEANS

1. General procedure. Suppose a company claims that the average collection times for the debts owing to it is 30 days. Liquidity is a problem to the company, and the company accountant suspects that debts are not collected in fact as rapidly as this. He wishes to test the assumption, and takes a sample of 40 invoices from last year's trading. The number of days outstanding is recorded, and the sample mean and variance are calculated as $\bar{x} = 34$ days, and

$$s^2 = \Sigma(x_i - \bar{x})^2/(n - 1) = 110 \text{ days.}$$

Two hypothesis are set up, called H_0 and H_1:

$$H_0: \mu = 30 \text{ days.}$$
$$H_1: \mu \neq 30 \text{ days.}$$

The *null hypothesis, H_0,* is so called because it involves no change from the supposed state of affairs. H_1 is referred to as the *alternative hypothesis*.

Our chain of reasoning begins with the assumption that H_0 is correct. If H_0 is true, then the sample mean is a random variable with a normal distribution, *i.e.*

$$\bar{x} \sim N(\mu, \sigma^2/n), \text{ or, approximately, } \bar{x} \sim N(\mu, s^2/n).$$

Thus:

$$\frac{\bar{x} - 30}{\sqrt{(s^2/n)}} \sim N(0, 1).$$

Now our particular sample value for the mean is 34, leading to a standardised normal variable of

$$(34 - 30)/\sqrt{(110/40)} = 2\cdot41.$$

The standard normal variable lies between $\pm 1\cdot96$ with a probability of 95 %, so on the assumption that H_0 is true the value of $2\cdot41$ is very unlikely to occur. We say that at the 5 % significance level H_0 can be rejected. In other words the logical consequences of assuming H_0 to be true lead us to an unlikely result, and therefore H_0 is discredited. Notice that we never say with absolute certainty that H_0 is untrue, but we can reject it in this case with probability $0\cdot05$ of making the wrong inference. If we had set our significance level at 1 %, the normal tables tell us that at this level the critical value would be $2\cdot57$. In other words if we required our error of incorrect rejection to be as small as 1 % then the sample evidence is insufficiently strong.

In summary, the logical sequence of events in performing this, and indeed any, statistical test of a hypothesis is:

(*i*) State null and alternative hypothesis, *e.g.*

$$H_0: \mu = 30.$$
$$H_1: \mu \neq 30.$$

(*ii*) Choose appropriate significance level, a, *e.g.*

$$a = 0\cdot05.$$

(*iii*) Choose appropriate test statistic, on the assumption that H_0 is true, *e.g.*

$$(\bar{x} - 30)/(s/\sqrt{n}).$$

(*iv*) Obtain critical values of test statistic which will determine acceptance or rejection of H_0, *e.g.*

Reject H_0 if $\bar{x} > 33\cdot25$, or $\bar{x} < 26\cdot75$.
Accept H_0 if $26\cdot75 \leqslant \bar{x} \leqslant 33\cdot25$.

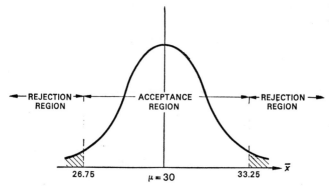

FIG. 52.—Acceptance/rejection regions for test of $H_0: \mu = 30$, $H_1: \mu \neq 30$.

These regions are referred to as the *rejection and acceptance regions* for the test (*see* Fig. 52).

(*v*) Calculate sample value of test statistic, *e.g.*

$$\bar{x} = 34.$$

(*vi*) Reject/accept H_0 depending on result of (*v*), *e.g.*

$\bar{x} = 34$ falls in the rejection region, so H_0 is rejected.

In developing our example the sequence was reordered slightly to assist in the explanation. The significance level (*ii*) should, however, be determined before taking the sample. As a general rule, it is useful to bear in mind that the null hypothesis is the one expected to be false, and the alternative hypothesis the one suspected to be true.

2. One-tail test for mean population.

(*a*) *Upper-tail test.* The accountant of our previous example may not be concerned with the possibility that mean debt-collection time is less than 30 days, since this would presumably be all to the good, and would not require any remedial action on his part. His interest is much more likely to be in the possibility that $\mu > 30$ days, since this may well have a serious effect on liquidity. A symbolic statement of his interest is supplied therefore by the hypotheses

(*i*) $$H_0: \mu = 30.$$
$$H_1: \mu > 30.$$

Following the same pattern as in Section 1 above:

(*ii*) Rejection of H_0 is to be followed by the introduction of a new scheme of preferential discounts in order to encourage prompt payments. It is decided that an acceptable level of the risk of unnecessarily introducing such a scheme is 0·02. Thus

$$a = 0·02.$$

(*iii*) The test statistic is

$$T = \frac{(\bar{x} - 30)}{\sqrt{s^2/n}}$$

and has a standard normal distribution if H_0 is true.

(*iv*) The probability is 0·02 that a standard normal variable *exceeds* the value 2·06. This, then, is the critical value for T. If T exceeds 2·06 then either (*a*) he has been unfortunate and—although H_0 is true—has selected an atypical sample, or (*b*) H_0 is untrue, and the mean of the population is greater than 30.

The rejection (of H_0) region is therefore defined by:

$$\frac{(\bar{x} - 30)}{\sqrt{(s^2/n)}} \geqslant 2·06$$

i.e.,

$$\bar{x} \geqslant 30 + 2·06 \sqrt{(s^2/n)}$$

and the acceptance (of H_0) region by:

$$\frac{(\bar{x} - 30)}{\sqrt{(s^2/n)}} < 2·06$$

i.e.,

$$\bar{x} < 30 + 2·06 \sqrt{(s^2/n)} = 33·42 \ (\textit{see Fig. 53}).$$

(*v*) Sample value of $\bar{x}$ is 34.
(*vi*) Reject H_0.

(*b*) *Lower-tail test.* Consider a proposed modification to the

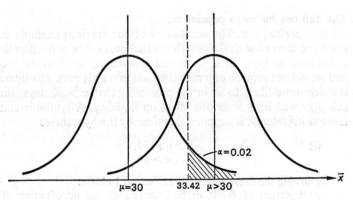

FIG. 53.—Test of $H_0: \mu = 30$ against $H_1: \mu > 30$.

system used by a company to record the numbers of hours worked by employees, their rates of pay, deductions, and so on. It is thought that the revised procedure will reduce the clerical time required to compute and record an employee's net pay from the current average of 8·5 minutes.

An appropriate test takes the following form:

(i)
$$H_0: \mu = 8\cdot5.$$
$$H_1: \mu < 8\cdot5.$$
(ii)
$$a = 0\cdot01.$$

(A high level of significance is chosen, since adoption of the new system will involve considerable reorganisation of the finance section's clerical staff, and the risk of doing this needlessly must be kept very small.)

(iii) Test statistic is

$$T = (\bar{x} - 8\cdot5)/\sqrt{(s^2/n)}$$

which has a standard normal distribution if H_0 is true.

(iv) Critical value for T is $-2\cdot33$. If $T < -2\cdot33$, H_0 is rejected, if $T > -2\cdot33$, H_0 is accepted (see Fig. 54). Rejection region is therefore $T < -2\cdot33$.

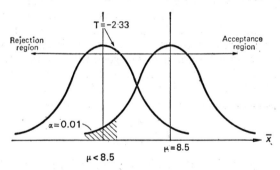

FIG. 54.—Test of $H_0: \mu = 8\cdot5$ against $H_1: \mu < 8\cdot5$.

A pilot scheme is implemented for a sample of 30 employees. Clerical time is found to average 7·8 minutes, with a standard deviation of 2·3 minutes.

Thus,

$$T = \frac{7\cdot8 - 8\cdot5}{\sqrt{(2\cdot3)^2/30}} = \frac{-0\cdot70}{0\cdot4199} = -1\cdot67.$$

This T-value falls in the acceptance region. There is insufficiently strong evidence to reject H_0.

Had an a-value of 0·05 been thought adequate, the critical

T-value for the test would have been −1·64, and the accountant would have (just) rejected H_0 at this level of significance.

3. Less specific tests concerning population means. In statistical testing there are two types of hypothesis—the *simple* and the *compound*. The first of these assumes a specific value for the parameter which is of interest. On the other hand, a compound hypothesis assumes a range of values for the parameter. In our earlier examples, all the null hypotheses were simple, and all the alternative hypotheses were compound. However, it is quite sensible to think in terms of a test involving two simple hypotheses,

e.g.

$$H_0: \mu = 30 \qquad (i)$$
$$H_1: \mu = 35$$

or, indeed, of two compound hypotheses,

e.g.

$$H_0: \mu \geqslant 30 \qquad (ii)$$
$$H_1: \mu < 30$$

An appropriate use of the second of these would arise if our accountant in Section 1 had no real idea of the average period for collection of debts. However, he has decided that 30 days is a desirable figure. Any greater than this is likely to cause his company liquidity problems, any less will result in loss of goodwill since many customers rely on trade credit as a source of short-term finance. He wishes to know in which situation the company finds itself. Accordingly he sets up the above hypotheses (*ii*). Choosing $a = 0.01$, and using the test statistic

$$T = \frac{\bar{x} - 30}{\sqrt{(s^2/n)}},$$

H_0 will be rejected if $T < -2.33$ (*i.e.*, $\bar{x} < 26.14$, *see* Fig. 55).

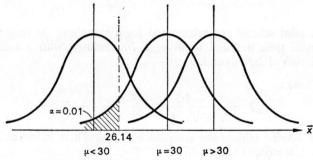

FIG. 55.—Test of $H_0: \mu \geqslant 30$ against $H_1: \mu < 30$.

The maximum probability of rejecting H_0 when true occurs when $\mu - 30$, and is then 1%. If $\mu > 30$, the probability of incorrectly rejecting H_0 is less than 1%.

4. Large and small sample tests.

The foregoing examples have invariably been concerned with tests involving large samples (*i.e.*, >20). This has been done deliberately in order that the logic and methodology of hypothesis testing could be explained in their simplest form. We cannot escape the fact, however, that samples are not always large. In fact, the modification which is made necessary to the previous analysis by the presence of small samples is a simple and natural one. Referring to Section 1 above, the only step in the argument that involves a specific probability distribution is (*iii*), where we use the fact that the sample mean has approximately a normal distribution with mean μ, variance s^2/n. Had the sample been small we should have had to use the result of the preceding chapter, *viz.* that

$$\frac{\bar{x} - 30}{\sqrt{(s^2/n)}} \sim t_{n-1}.$$

Taking up the argument at this point, and supposing for definiteness that the sample size had been 15:

(*vi*) Critical value of

$$\frac{\bar{x} - 30}{\sqrt{(s^2/n)}} \text{ is } 2\cdot145,$$

since 95% of the distribution of t_{14} lies between $\pm 2\cdot145$.

Thus, reject H_0 if $\qquad \bar{x} > 30 + 2\cdot145 \sqrt{(110/15)} = 35\cdot81$
$\qquad\qquad$ or $\qquad \bar{x} < 30 - 2\cdot145 \sqrt{(110/15)} = 24\cdot19$
and accept H_0 if $\qquad\qquad 24\cdot19 < \bar{x} < 35\cdot91.$

Since $\bar{x} = 34$, H_0 is accepted.

Notice that the acceptance region here is wider than in Section 1 above. The reason lies in the fact that, although we have assumed sample means of 34, and sample variances of 110 in each case, less reliability can be attached to these values when they arise from a sample of 15 rather than 40. Our line of argument is that H_0 is to be accepted unless evidence is presented to the contrary. When derived from a small sample only, $\bar{x} = 34$ (and $s^2 = 110$) forms evidence of insufficient quality to enable us to reject H_0.

An examination of the other tests involving the mean which have been introduced will reveal that a similar modification to stage (*iii*) of the argument in each instance will render the test suitable for use with small samples.

Worked Example

The personnel department of a company developed an aptitude test for screening potential employees. The person devising the test asserted that the mean mark attained would be 100. The following results were obtained with a random sample of applicants:

$$\bar{x} = 96, \ s = 5.2, \ n = 13.$$

Perform a test of this hypothesis against the alternative that the mean mark is less than 100, at the 5% significance level.

Answer

(i) $H_0: \mu = 100.$
                    $H_1: \mu < 100.$

(ii) $a = 0.05.$

(iii) Test statistic: $T = \dfrac{\bar{x} - \mu}{\sqrt{(s^2/n)}} \sim t_{n-1}.$

(iv) Critical value of $T = -1.782$ (see Fig. 56).
     Rejection region is $\bar{x} < 100 - 1.782 \sqrt{(5.2^2/13)} = 97.43.$

(v) Sample value $= 96.$

(vi) Reject H_0. Since the sample value of $\bar{x}$ is only just significant at 5% level, it would be wise to perform a second test with another sample.

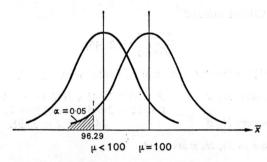

FIG. 56.—Test of $H_0: \mu = 100$ against $H_1: \mu < 100$.

Exercises on Section 4

1. A package-filling device is set to fill cereal boxes with a mean weight of 20·00 ounces. The standard deviation is known to be 0·5 ounces. It is suspected that a filler is overfilling the boxes, thereby increasing materials costs. A random sample of 25 filled boxes is weighed (net weight), and has a mean of 20·20. What conclusion can you draw?

2. Using a standard manufacturing process, the breaking strength of copper wire is a random variable with mean 1000 and standard deviation 5·4. A new, cheaper process is tested—a sample of 28 wires is drawn with $\bar{x} = 984$, $s = 6.1$. The manufacturer requires to be, in his own words,

"95% sure" that the new process will produce wire as strong as the old process. Formulate and perform an appropriate test.

3. A manufacturer claims that the average life of a grinding wheel is 100 hours, with a standard deviation of 20 hours (which is not in question). A customer tests 100 wheels. Set up a rule which would reject the manufacturer's claim only 5% of the time if it is true.

4. Absentee rates per thousand for a factory over the last 12 months have been:

64	74
72	67
79	49
58	52
75	40
83	46

The unions claim that absentee rates average 58 per thousand. Design and perform an appropriate test of this claim.

5. Shovel and Pick Ltd. have over the last twenty-five days had the following output figures:

125	300	275	250	225
250	325	175	150	175
212	112	165	300	405
375	350	360	374	270
212	178	232	254	185

The management claim an average level of output of 290/day. Do the results support their claim?

TWO SAMPLE TESTS INVOLVING POPULATION MEANS

5. Test for equality of population means. In sampling from two populations, as in Chapter 7, we determined that the statistic

$$T = \frac{(\bar{x}_1 - \bar{x}_2) - (\mu_1 - \mu_2)}{\sqrt{\left\{\frac{s_1^2}{n_1} + \frac{s_2^2}{n_2}\right\}}}$$

has approximately a standard normal distribution when n_1 and n_2 are large. Although no similar comparatively simple result obtains when the samples are small, under the conditions that the populations are approximately normal and have the same variance we have

$$T = \frac{(\bar{x}_1 - \bar{x}_2) - (\mu_1 - \mu_2)}{\sqrt{\left\{s^2\left(\frac{1}{n_1} + \frac{1}{n_2}\right)\right\}}} \sim t_{n_1+n_2-2}$$

where s^2 is the pooled variance.

Knowledge of the distributions of these two statistics enables us to test hypotheses concerning the relative values of the population means. No doubt because of his sterling work on the collection of debts, the accountant of Section 1 has recently been made group accountant. He is interested now in determining whether the average collection period varies from one division of the company to another, and gathers the following information:

Division I	Division II
$\bar{x}_1 = 36$	$\bar{x}_2 = 41$
$s_1^2 = 129$	$s_2^2 = 148$
$n_1 = 37$	$n_2 = 32$

(a) First let us suppose that he is interested only in whether the means differ. He proceeds as follows:

(i) $\qquad\qquad H_0: \mu_1 = \mu_2.$
$\qquad\qquad\qquad H_1: \mu_1 \neq \mu_2.$
(ii) $\qquad\qquad a = 0.05.$

(iii) Test statistic: $T = \dfrac{(\bar{x}_1 - \bar{x}_2)}{\sqrt{\left(\dfrac{s_1^2}{n_1} + \dfrac{s_2^2}{n_2}\right)}} \sim N(0, 1)$

since $\mu_1 - \mu_2 = 0$ if H_0 is true.

(iv) Critical values of T: ± 1.96.
Reject H_0 if $T < -1.96$, or $T > 1.96$.

A two-tailed test is used here, since evidence that either $\mu_1 > \mu_2$, or that $\mu_2 > \mu_1$ is equally relevant in discrediting H_0.

(v) Sample results: $T = (36 - 41) \Big/ \sqrt{\left(\dfrac{129}{37} + \dfrac{148}{32}\right)} = -1.756$.

(vi) H_0 is accepted.

(b) Second, suppose that our accountant suspected that $\mu_1 > \mu_2$; the appropriate hypotheses would then be:

$$H_0: \mu_1 = \mu_2.$$
$$H_1: \mu_1 > \mu_2.$$

Using the same level of significance, and the same test statistic T, the critical value will be 1.64, since an upper-tailed test is required. The critical region is now $T > 1.64$, and since the sample value is -1.756, H_0 cannot be rejected.

(c) Finally, if his suspicion was that $\mu_1 < \mu_2$,

then:

$$H_0: \mu_1 = \mu_2$$
$$H_1: \mu_1 < \mu_2$$

and a lower-tail test with $a = 0.05$ has a rejection region of

$$T < -1.64.$$

In this case H_0 would be rejected in favour of H_1.

The final test may appear at first sight to contradict the first and second, since they both result in acceptance of H_0. However, one has to bear in mind that we are asking for a choice between different pairs of alternatives in each case, and that H_0 is "innocent" until proved "guilty" (*i.e.*, accepted in the absence of sufficient evidence to contradict it). Furthermore, remember that tests are constructed *before* drawing the sample, and the fact that different alternative hypotheses are used in (*i*)–(*iii*) indicates that different *a priori* beliefs in the relative values of μ_1, μ_2 are held in each case. An analogy may be helpful. Suppose you are interested in going to one of two new detective films *Shift* and *The French Contraption*. You read the reviews of the two films which may lead you to form one of three *a priori* beliefs:

  (*i*) That there is very little to choose between them.
  (*ii*) That *Contraption* is preferable to *Shift*.
  (*iii*) That *Shift* is preferable to *Contraption*.

In cases (*i*) and (*ii*) the subsequent advice of a friend that *Shift* is "somewhat better" than *Contraption* is unlikely to be sufficient evidence for you to decide that they are not equally good. However, the same advice in the context of your *a priori* belief (*iii*), could, since it reinforces that belief, enable you to reject the hypothesis that they are equally good.

6. Test for difference between means. The following test is a straightforward extension of the previous one, and has as its null hypothesis one of the form:

$$H_0: \mu_1 - \mu_2 = \mu_0, \text{ say.}$$

It tests, therefore, for a specific non-zero difference between the population means, rather than for zero difference (*i.e.*, equality). The alternative hypothesis may, for example, be one of

$$H_1: \mu_1 - \mu_2 \neq \mu_0.$$
$$H_1: \mu_1 - \mu_2 > \mu_0.$$

or

$$H_1: \mu_1 - \mu_2 < \mu_0.$$

The test statistics appropriate are simply those given at the beginning of Section 5 above, *i.e.* on the assumption that H_0 is true:

$$T = \frac{(\bar{x}_1 - \bar{x}_2) - \mu_0}{\sqrt{\left\{\dfrac{s_1^2}{n_1} + \dfrac{s_2^2}{n_2}\right\}}} \sim N(0, 1)$$

in the large sample case, and

$$T = \frac{(\bar{x}_1 - \bar{x}_2) - \mu_0}{\sqrt{\left\{ s^2 \left(\frac{1}{n_1} + \frac{1}{n_2} \right) \right\}}} \sim t_{n_1 + n_2 - 2}$$

for small samples.

Worked Examples

1. A bank is considering opening a new branch in one of two neighbourhoods. It desires to open the branch in the neighbourhood having the higher mean income. From census records, two random samples of sizes 15 and 20 are selected, and the following information obtained:

$$n_1 = 15 \qquad n_2 = 20$$
$$\bar{x}_1 = £3000 \qquad \bar{x}_2 = £3100$$
$$\sqrt{\left\{ \frac{\Sigma(x_{i1} - \bar{x}_1)^2}{n_1 - 1} \right\}} = s_1 = £150 \qquad s_2 = £200 = \sqrt{\left\{ \frac{\Sigma(x_{i2} - \bar{x}_2)^2}{n_2 - 1} \right\}}.$$

The first neighbourhood also has several small businesses that might be attracted to a new branch. The bank, therefore, wishes to avoid the error of concluding that the second neighbourhood has a higher mean income, when the true state of nature is that the first neighbourhood has a mean income equal to or greater than the second neighbourhood. Set up the appropriate hypotheses and test at the 5% level of significance.

Answer

    (*i*) $H_0: \mu_1 \geqslant \mu_2.$
                $H_1: \mu_1 < \mu_2.$

    (*ii*) Since the error of incorrectly rejecting H_0 is to be small, take

$$a = 0 \cdot 01.$$

    (*iii*) Test statistic: $T = \dfrac{(\bar{x}_1 - \bar{x}_2)}{\sqrt{\left\{ s^2 \left(\dfrac{1}{n_1} + \dfrac{1}{n_2} \right) \right\}}} \sim t_{n_1 + n_2 - 2}$

(we assume population variances are equal, and calculate the pooled variance, s^2.)

    (*iv*) Critical value of T is $-2 \cdot 442$. A lower-tail test is necessary, since a large negative value of T indicates departure from the null hypothesis that $\mu_1 - \mu_2 \geqslant 0$.

    (*v*) Sample results:

$$s^2 = \frac{(14 \times 150) + (19 \times 200)}{15 + 20 - 2}$$
$$= 178 \cdot 79.$$

Thus,

$$T = \frac{(3000 - 3100)}{\sqrt{\left\{ 178 \cdot 79 \left(\frac{1}{15} + \frac{1}{20} \right) \right\}}}$$
$$= -21 \cdot 9.$$

    (*vi*) Reject H_0, since $-21 \cdot 9 < -2 \cdot 44$ (*see* Fig. 57).

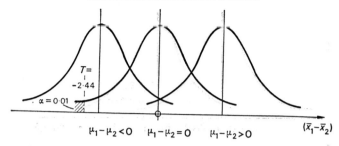

FIG. 57.—Test of $H_0: \mu_1 \geq \mu_2$ against $H_1: \mu_1 < \mu_2$ (small sample).

2. A dispute is fomenting between the malters and yeasters at the Hopless brewery. Malters claim that their average take-home pay, including over-time, is £3·20 less than that of the yeasters.

As company accountant you suspect this claim to be true. Design an appropriate test of the malters' claim, and implement your test using the following sample information taken from the finance section's recent records:

Malters	Yeasters
$n_1 = 50$	$n_2 = 40$
$\bar{x}_1 = £30·25$	$\bar{x}_2 = £35·40$
$s_1 = £\sqrt{19·85}$	$s_2 = £\sqrt{17·62}$

Answer

(i) $H_0: \mu_2 - \mu_1 = 3·20.$
    $H_1: \mu_2 - \mu_1 > 3·20.$

(ii) $a = 0·01.$

(iii) Test statistic: $T = \dfrac{(\bar{x}_2 - \bar{x}_1) - (\mu_2 - \mu_1)}{\sqrt{\left(\dfrac{s_1^2}{n_1} + \dfrac{s_2^2}{n_2}\right)}} \sim N(0, 1).$

(iv) Critical value of T is 2·33. An upper tail-test is appropriate since large positive values of T discredit H_0 in favour of H_1.

(v) Sample results:

$$T = \frac{(35·40 - 30·25) - 3·20}{\sqrt{\left(\dfrac{19·85}{50} + \dfrac{17·62}{40}\right)}}$$
$$= 2·13.$$

(vi) H_0 cannot be rejected (*see* Fig. 58).

Exercises on Section 6

6. In a certain year, the mean interest rate on loans to all large retailers was $6·0\%$ and the standard deviation $0·2\%$. Two years later, a random sample of 100 loans to large retailers had a mean rate of interest of $6·015\%$

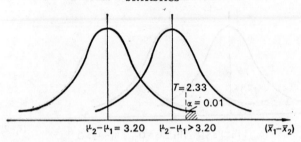

FIG. 58.—Test of H_0: $\mu_2 - \mu_1 = 3\cdot20$ against H_1: $\mu_2 - \mu_1 > 3\cdot20$.

with the same standard deviation. Would you be willing to conclude that there has been a change in the average rate? Assume you are willing to run a 5% risk of concluding that there has been a change when in fact there was none.

7. A portfolio manager is running two portfolios of stocks. The average values of rates of return on the two portfolios are given—with their standard deviations—in the following table.

$$\textit{Portfolio I} \qquad\qquad \textit{Portfolio II}$$
$$n_1 = 12 \qquad\qquad n_2 = 15$$
$$\bar{x}_1 = 13\% \qquad\qquad \bar{x}_2 = 9\cdot5\%$$
$$\sqrt{\left\{\frac{\Sigma(x_{i1} - \bar{x}_1)^2}{n_1 - 1}\right\}} = s_1 = 4\% \qquad \sqrt{\left\{\frac{\Sigma(x_{i2} - \bar{x}_2)^2}{n_2 - 1}\right\}} = s_2 = 3\%$$

where n_1 and n_2 are the numbers of quarterly periods over which the averages are taken. Is his belief that portfolio I performs better than portfolio II justified?

8. For digesters of a certain shape and size, used in paper-making, the mean weight of fuel per day is 125·1 units. A new shape of digester of the same capacity when tested over 10 days used the following weights of fuel:

125·15	124·84	123·32	122·50	125·32
124·57	123·54	127·43	124·82	124·41

Has the new shape made an improvement in fuel economy?

9. The paper currency used by the country of Lilliput is the puttee. As a professional counterfeiter you are employed by the government of the neighbouring country of Brobdignag to introduce forged puttees into Lilliput with the object of undermining their economy. As a test of the successfulness of your forged notes a number were introduced into Lilliput some time ago. Seven large cash withdrawals are made from various secret accounts in Lilliputian banks, and the percentage of forged notes in each withdrawal is as follows:

0·0363, 0·0356, 0·0343, 0·0358, 0·0362, 0·0374, 0·0365.

After a further few weeks a second series of withdrawals are made with the results:

0·0340, 0·0355, 0·0351, 0·0354, 0·0363, 0·0365, 0·0356.

Are you justified in assuming that your counterfeit notes have been absorbed undetected into the Lilliputian circulation?

7. Tests between population means: two dependent samples.

We have implicitly assumed hitherto that the comparison of two populations has been on the basis of independent samples. Indeed, it is this assumption alone that has enabled us to derive the variance of the sampling distribution of $(\bar{x}_1 - \bar{x}_2)$ as the sum of the variances of $\bar{x}_1$ and $\bar{x}_2$,

i.e.

$$\left(\frac{s_1^2}{n_1} + \frac{s_2^2}{n_2} \right).$$

The assumption, and the sampling distribution of T which follows, are not valid when the samples are dependent. A further test—known as the *paired comparison test*—has been developed for dealing with situations where there is a natural pairing of sampled values. We may, for instance, have sets of budgeted and actual figures by departments for a particular year, productivity figures for a set of factories before and after the introduction of a new wage incentive scheme, or measures of the performance of a number of patients before and after their treatment with a new drug. In each case the samples are dependent in the sense that each department, factory, or patient gives rise to a figure in the first sample, and one in the second.

Consider the first example, of budgeted and actual costs, by departments (*see* Table 26).

To determine whether there is any overall significant difference between budgeted and actual costs, we first form the differences

Table 26. Department sales—budgeted and actual

Department	Budgeted Sales (£000)	Actual Sales (£000)	Difference d	d^2
A	2·5	3·2	+0·7	0·49
B	5·1	7·6	+2·5	6·25
C	7·3	6·3	−1·0	1·00
D	6·4	8·9	+2·5	6·25
E	3·9	4·0	+0·1	0·01
F	8·7	9·7	+1·0	1·00
G	9·5	7·1	−2·4	5·76
H	4·0	2·3	−1·7	2·89
I	6·8	8·6	+1·8	3·24
J	11·3	10·1	−1·2	1·44
K	10·9	13·8	+2·9	8·41
L	9·7	12·7	+3·0	9·00
		Total	8·2	45·74

(taking account of sign), d, between the pairs of figures. Conceptually this set of differences can be considered as a random sample from the population of differences which might have occurred. Furthermore, on the reasonable assumption that this population has a more or less normal distribution, we can state the null and alternative hypotheses:

(i) $H_0: D = 0$
                                       $H_1: D \neq 0$

where D is the mean of the population of differences.

(ii) Let $a = 0 \cdot 10$

(iii) The statistic: $T = \dfrac{\bar{d} - D}{\sqrt{(s^2/n)}}$

where s = standard deviation of d-values

$$= \sqrt{\left\{ \frac{\Sigma d^2 - n\bar{d}^2}{n - 1} \right\}}$$

will have a t-distribution with $(n - 1)$ degrees of freedom (cf. Section 4 above). Under the null hypothesis $(D = 0)$,

$$\bar{d}/\sqrt{\{s^2/n\}} \sim t_{n-1}.$$

(iv) Critical values for T are $\pm 1 \cdot 80$, since H_1 is a two-sided alternative hypothesis.

(v) Sample results: $T = \dfrac{\{(8 \cdot 2/12) - 0\} \sqrt{12}}{1 \cdot 9102} = 1 \cdot 239.$

(vi) H_0 cannot be rejected, since T falls in the acceptance region. There is no significant difference between budgeted and actual costs at the 10% significance level.

Exercises on Section 7

10. As an accountant you are presented with the following data for budget variances:

Budget	1977	1976	Budget	1977	1976
A	1020	900	F	900	940
B	850	870	G	540	510
C	790	690	H	610	540
D	1060	860	I	710	650
E	750	730	J	970	790

(i) Is there any reason to believe that achievements against budgets are slipping?

(ii) Construct a 95% confidence interval for the true difference.

11. Mammon Ltd. is faced with the choice between two five-year projects, A and B. The cash flows for two similar projects over the past five years have been:

	0	1	2	3	4	5
A	−1000	500	−600	700	300	100
B	−900	550	−500	800	200	50

Conduct a test to show if there is any significant difference between the two sets of cash flows from the two projects in terms of their present value. Assume a discount rate of 8% per annum.

12. Breakdown costs of two machines, the Castor and the Pollux, over the past 6 months have been:

	J	F	M	A	M	J
Castor	6·6	7·0	7·7	5·6	5·1	4·2
Pollux	7·3	8·5	7·6	6·9	5·4	4·4

Is there any reason to suppose that one machine is more costly (in terms of breakdowns) than the other?

TESTS CONCERNING PROPORTIONS

8. Large samples. Remembering that for a large sample the sampling proportion r/n has approximately a normal distribution with mean p and variance $p(1 - p)/n$ (*see* Chapter 7, Section 14), it is a straightforward matter to construct tests concerning the population proportion p, based on the test statistic:

$$T = \left(\frac{r}{n} - p\right) \Big/ \sqrt{\left\{\frac{p(1 - p)}{n}\right\}} \sim N(0, 1).$$

Take the case of Dazzle Co. Ltd., which is contemplating the marketing of a soap powder, containing an innovatory pathological whitener, and to be called Pathoblanc. Two samples of housewives are invited to try the new powder, and to indicate subsequently whether they would use it if it appeared on the market. One sample is of current users of Dazzle's present product Bioblanc. The other sample is of non-users of Bioblanc. Results are as follows:

	No. who would change to Pathoblanc	No. who would not change to Pathoblanc	
No. who currently use Bioblanc	73 (r_1)	119	192 (n_1)
No. who do not currently use Bioblanc	186 (r_2)	266	452 (n_2)
	259	385	644

It is hoped that brand loyalty will result in at least 30% of Bioblanc users changing to Pathoblanc. To test this supposition Dazzle's statistician proceeds thus:

(i) $$H_0: p_1 = 0.30$$
$$H_1: p_1 > 0.30$$

(ii) $$a = 0.025$$

(iii) Test statistic: $$T = \frac{\left(\dfrac{r_1}{n_1} - 0.30\right)}{\sqrt{\left\{\dfrac{r_1}{n_1}\left(1 - \dfrac{r_1}{n_1}\right)\Big/ n_1\right\}}} \sim N(0, 1)$$

if H_0 is true.

(iv) An upper-tail test is appropriate since a large positive value of T discredits H_0 in favour of H_1. The critical value for T is therefore 1.96, and the rejection region is defined by $T > 1.96$.

(v) Sample results: $$T = \frac{\left(\dfrac{73}{192} - 0.30\right)}{\sqrt{\left\{\dfrac{73}{192}\left(1 - \dfrac{73}{192}\right)\Big/ 192\right\}}} = 2.2898.$$

(vi) H_0 is rejected in favour of H_1.

Second, the statistician wishes to test the presumption that the proportions of users and non-users of Bioblanc who would change to Pathoblanc differ. This requires a little more thought. We know that:

$$\frac{r_1}{n_1} \sim N\left(p_1, \frac{p_1(1 - p_1)}{n_1}\right)$$

and

$$\frac{r_2}{n_2} \sim N\left(p_2, \frac{p_2(1 - p_2)}{n_2}\right).$$

So

$$\frac{r_1}{n_1} - \frac{r_2}{n_2} \sim N\left\{(p_1 - p_2), \frac{p_1(1 - p_1)}{n_1} + \frac{p_2(1 - p_2)}{n_2}\right\}$$

i.e.

$$\frac{\left(\dfrac{r_1}{n_1} - \dfrac{r_2}{n_2}\right) - (p_1 - p_2)}{\sqrt{\left\{\dfrac{p_1(1 - p_1)}{n_1} + \dfrac{p_2(1 - p_2)}{n_2}\right\}}} \sim N(0, 1).$$

The statistician's present null hypothesis is that $p_1 = p_2$, or $p_1 - p_2 = 0$. Accordingly, his test statistic ideally would be:

$$\frac{\left(\dfrac{r_1}{n_1} - \dfrac{r_2}{n_2}\right)}{\sqrt{\left\{p(1-p)\left(\dfrac{1}{n_1} + \dfrac{1}{n_2}\right)\right\}}} \sim N(0, 1).$$

However, the assumed common proportion p is not known of course. As a better estimate of p than is provided by r_1/n_1 or r_2/n_2 alone he may use

$$\frac{r_1 + r_2}{n_1 + n_2} = \frac{r}{n}.$$

Finally, then, the appropriate test statistic having an approximately standard normal distribution is

$$T = \frac{\left(\dfrac{r_1}{n_1} - \dfrac{r_2}{n_2}\right)}{\sqrt{\left\{\dfrac{r}{n}\left(1 - \dfrac{r}{n}\right)\left(\dfrac{1}{n_1} + \dfrac{1}{n_2}\right)\right\}}}.$$

His test now proceeds along the lines:

(i) $$H_0: p_1 = p_2$$
$$H_1: p_1 \neq p_2$$

(ii) $$a = 0{\cdot}025.$$

(iii) Test statistic above.

(iv) A two-tailed test with critical values for T of $\pm 2{\cdot}24$ is needed. The rejection region is therefore defined by $T > 2{\cdot}24$ and $T < -2{\cdot}24$.

(v) Sample results:

$$T = \frac{\left(\dfrac{73}{192} - \dfrac{186}{452}\right)}{\sqrt{\left\{\dfrac{259}{644}\left(1 - \dfrac{259}{644}\right)\left(\dfrac{1}{192} + \dfrac{1}{452}\right)\right\}}} = -0{\cdot}7413.$$

(vi) H_0 is accepted since $-0{\cdot}7413$ is not in the rejection region.

9. Small samples. If only a small sample is available for estimating a population proportion the normal approximation which has been used in Section 8 above is no longer applicable. Instead we have to turn to the exact sampling distribution of r, which is of course binomial.

As an illustration of the approach, consider a sample of 10 advertisers in the "Accounting Weakly" who are asked by the editors of the magazine whether they were or were not satisfied with the response to their advertisements. 8 claim they were satisfied, 2 do not. The editors had assumed a satisfaction rate of 60% although they did hope for better. They require to test their assumption at the 10% significance level.

(i) $$H_0: p = 0.60.$$
$$H_1: p > 0.60.$$
(ii) $$a = 0.10.$$

(iii) The test statistic is the number of satisfied customers, r, in the sample. If H_0 is true, then in a sample of 10 the probabilities of having just r satisfied customers are:

$$p(r = 0) = \frac{10!}{0! \, 10!} (0.60)^0 (0.40)^{10} = 0.0001$$

$$p(r = 1) = 0.00157$$
$$p(r = 2) = 0.01062$$
$$p(r = 3) = 0.04247$$
$$p(r = 4) = 0.11148$$
$$p(r = 5) = 0.20066$$
$$p(r = 6) = 0.25082$$
$$p(r = 7) = 0.21499$$
$$p(r = 8) = 0.12093$$
$$p(r = 9) = 0.04031$$
$$p(r = 10) = 0.00605$$

The cumulative probabilities are therefore:

$$p(r \leqslant 0) = 0.00010$$
$$p(r \leqslant 1) = 0.00167$$
$$p(r \leqslant 2) = 0.01229$$
$$p(r \leqslant 3) = 0.05476$$
$$p(r \leqslant 4) = 0.16624$$
$$p(r \leqslant 5) = 0.36690$$
$$p(r \leqslant 6) = 0.61772$$
$$p(r \leqslant 7) = 0.83271$$
$$p(r \leqslant 8) = 0.95364$$
$$p(r \leqslant 9) = 0.99395$$
$$p(r \leqslant 10) = 1.00000$$

The distribution of the test statistic, under the assumption that H_0 is true, is therefore specified.

(iv) An upper-tail test is appropriate, and we note that the probability of obtaining a value of 9 or more if $p = 0.60$ is 0.04636, and of 8 or more is 0.16729. The critical value of the test statistic at the 10% significance level (and, as a matter of fact, at the 5% level) is 9.

(v) Sample results: $r = 8$.

(vi) H_0 cannot be rejected.

Exercises on Section 9

13. The management of a grocery chain is considering whether or not to build a new store in an existing shopping centre. The manager of the shopping centre claims that more than 55% of all families living within a two-mile radius of the centre shop there. If the grocery chain management accepts the "greater than 55%" estimate as fact, the store will be built. If it is concluded that the claim is too high, the store will not be built. A sample of 225 families is taken and, of these, 110 would use the new store. Would you feel that the management was justified in building the store?

14. The managing director of a company is concerned about the number of complaints which he receives attributed to "the computer." He counts these complaints and estimates that they cover 20% of the computer transactions. Since the computer company assure him that their accuracy is better than 99·9% he is suspicious. He therefore instructs his internal auditor to submit 10 different transactions to the system, and let him know how many are correctly performed. The auditor produces 9 correct results out of 10.

(*i*) What does this show the managing director?

(*ii*) Could you devise a better test?

15. A bungle consists of three components, tips T_i, tops T_o, and taps T_a. If any one of these components is defective then the bungle is inoperable. Renowned Bunglers Ltd. claim that no more than 10% of their bungles are defective. A sample of 1600 of their finished bungles is tested with the following results:

$$N(T_i) = 110 \qquad N(T_o) = 109 \qquad N(T_a) = 115$$
$$N(T_i \cap T_o) = 46 \qquad N(T_i \cap T_a) = 49 \qquad N(T_o \cap T_a) = 66$$
$$N(T_i \cap T_o \cap T_a) = 30$$

(here $N(T_i \cap T_a)$ represents the number of bungles with defective tips and taps, etc.).

Test whether the manufacturer's claim is justified or not.

16. A random sample of 400 accounts is examined—205 are in order, 195 are not. Let p be the proportion of satisfactory accounts in the large population from which the sample has been drawn.

Consider the hypotheses

$$H_0: p \leqslant 0\cdot5.$$
$$H_1: p > 0\cdot5.$$

H_0 is to be rejected if the sample proportion is greater than 0·54. What is the greatest probability of incorrectly rejecting H_0?

Design a test in which the probability of incorrectly rejecting H_0 is at most 0·01, and test the hypothesis using the sample data.

17. In the illustration concerning Bioblanc and Pathoblanc (Section 8 above) test the null hypothesis

$$H_0: p_2 = 0\cdot30$$

against the alternative

$$H_1: p_2 > 0\cdot30$$

where p_2 is the proportion of non-users of Bioblanc who will use Pathoblanc.

FINITE POPULATION CORRECTION

In common with our previous discussion of estimation we have assumed so far that our sample is not an appreciable fraction of the population (*see* Chapter 7, Section 15). When this is not the case, introduction of the finite population correction $\sqrt{\{(N - n)/(N - 1)\}}$ in the expression for standard deviation provides the necessary modification to the test statistics.

Worked Example

In the illustration of Section 6 (worked example 2), concerning Malters and Yeasters, suppose that the samples had been respectively drawn from populations of 200 and 180. Conduct the previous test with appropriate modification.

Answer

(i)
$$H_0: \mu_2 - \mu_1 = 3 \cdot 20.$$
$$H_1: \mu_2 - \mu_1 > 3 \cdot 20.$$

(ii)
$$a = 0 \cdot 05.$$

(iii) Using finite population correction,

$$\frac{s_1^2}{n_1}\left(\frac{N_1 - n_1}{N_1 - 1}\right) \qquad \text{replaces} \quad \frac{s_1^2}{n_1}$$

and

$$\frac{s_2^2}{n_2}\left(\frac{N_2 - n_2}{N_2 - 1}\right) \qquad \text{replaces} \quad \frac{s_2^2}{n_2}.$$

Modified test statistic is:

$$T = \frac{(\bar{x}_2 - \bar{x}_1) - (\mu_2 - \mu_1)}{\sqrt{\left\{\frac{s_1^2}{n_1}\left(\frac{N_1 - n_1}{N_1 - 1}\right) + \frac{s_2^2}{n_2}\left(\frac{N_2 - n_2}{N_2 - 1}\right)\right\}}} \sim N(0, 1).$$

(iv) Critical T value is $1 \cdot 645$.

(v) Sample results:

$$T = \frac{1 \cdot 95}{\sqrt{\left\{\frac{19 \cdot 85}{50}\left(\frac{200 - 50}{200 - 1}\right) + \frac{17 \cdot 62}{40}\left(\frac{180 - 40}{180 - 1}\right)\right\}}} = 2 \cdot 43.$$

(vi) This result is significant at the 1 % level, and H_0 is rejected. (Notice that the information concerning population sizes has led to the rejection of the previously accepted null hypothesis.)

ERRORS IN STATISTICAL HYPOTHESIS TESTING

10. Introduction. We have repeatedly referred to the probability of rejecting the null hypothesis H_0 when in fact it is true. This source of error is completely under our control when we design a statistical test, and is of course our chosen significance level, a. More technically, however, this error is known as *Type I error*, to distinguish it from a second source, the *Type II error*. The latter, usually denoted by β, is the probability of accepting H_0 when in fact it is false. The state of affairs surrounding any statistical test of a null against an alternative hypothesis is summarised in Table 27.

Table 27. Summary of hypothesis testing

Decision	Possible states of nature	
	H_0 true H_1 false	H_0 false H_1 true
Accept H_0, reject H_1	Correct decision (probability: $(1 - a)$)	Incorrect decision (probability: β)
Reject H_0, accept H_1	Incorrect decision (probability: a, *i.e.* significance level)	Correct decision (probability: $(1 - \beta)$)

11. Power functions and operating characteristic curves. To clarify this with respect to a particular example let us return to the accountant of Section 1, and let us suppose that the true value of the population mean is 32 days. Our decision rule was to accept

$$H_0 \text{ if } 26 \cdot 75 < \bar{x} < 33 \cdot 25.$$

The probability then, that H_0 will be accepted when in fact $\mu = 32$, is just the probability that a random sample drawn from a population with mean 32 will have a sample mean in the range 26·75 to 33·25. Thus, when $\mu = 32$,

$$\beta = P(26 \cdot 75 < \bar{x} < 33 \cdot 25)$$
$$= \Phi\left(\frac{33 \cdot 25 - 32}{1 \cdot 658}\right) - \Phi\left(\frac{26 \cdot 75 - 32}{1 \cdot 658}\right)$$
$$= \Phi(0 \cdot 7539) - \Phi(-3 \cdot 1665)$$
$$= 0 \cdot 7752.$$

This situation is shown graphically in Fig. 59.

In a similar way we can calculate the Type II (*i.e.*, β) error for the test if the true value of μ is 31, 29, or indeed any other value. The results of this exercise are presented in what is known as the *operating characteristic curve* for the test (*see* Fig. 60).

Alternatively, the performance of the test can be presented with the *power function*, defined by $(1 - \beta)$, on the vertical axis, and with the same horizontal axis. The resulting *power curve* is, of course, simply the operating characteristic curve inverted (*see* Fig. 61).

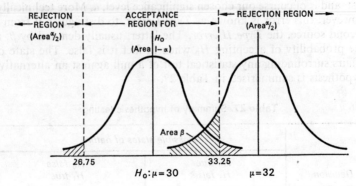

Fig. 59.—Type II error in test $H_0: \mu = 30$ against $H_1: \mu \neq 30$, when in fact $\mu = 32$.

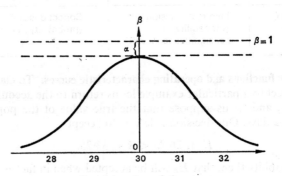

Fig. 60.—Operating characteristic curve for test $H_0: \mu = 30$, $H_1: \mu \neq 30$.

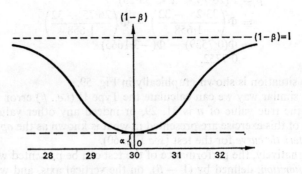

Fig. 61.—Power curve for test $H_0: \mu = 30$, $H_1: \mu \neq 30$.

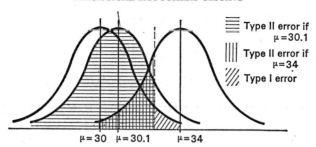

Fig. 62.—Type II error of test $H_0: \mu = 30$, $H_1: \mu \neq 30$ for true values of μ of 30·1 and 34.

As one might expect the probability, β, of accepting H_0 when it is false is greatest for true values of the population parameter near 30. If, for example, $\mu = 30·1$, so that H_0 was untrue, a sample would typically not carry sufficient information for us to reject $H_0: \mu = 30$. As the true population mean becomes progressively larger or smaller than the hypothesiscd valuc of 30, so the Type II error decreases (*see also* Fig. 62).

When dealing with an upper- or lower-tail test the power function and operating characteristic curve can be derived in a similar way. Taking our accountant's tests of Section 2 we suppose, for purposes of comparison with the two-tailed test, that his significance levels were again chosen to be 5 %. For his one-tailed tests we calculate the Type II errors when the true value of μ is (*i*) 31, and (*ii*) 29:

Upper-tail test (*see* Fig. 63).

$$H_0: \mu = 30.$$
$$H_1: \mu > 30.$$

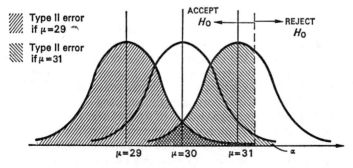

Fig. 63.—Type II error of test $H_0: \mu = 30$, $H_1: \mu > 30$.

Since $a = 0.05$, the critical value for

$$\frac{(\bar{x} - 30)}{\sqrt{(s^2/n)}} \text{ is } 1.64.$$

Thus H_0 will be rejected if $\bar{x} > 30 + 1.64\sqrt{(110/40)} = 32.72$.

(i) If μ is 31, the probability of accepting H_0 is just the probability of drawing a sample with a mean of less than 32.72 from a population with $\mu = 31$, i.e.

$$\beta = \Phi\left(\frac{32.72 - 31.0}{1.658}\right)$$
$$= \Phi(1.0373)$$
$$= 0.8502.$$

(ii) If μ is 29,

$$\beta = 1 - \Phi\left(\frac{32.72 - 29.00}{1.658}\right)$$
$$= 1 - \Phi(-2.2437)$$
$$= 0.9876.$$

Lower-tail test

$$H_0: \mu = 30.$$
$$H_1: \mu < 30.$$

With $a = 0.05$, rejection region for H_0 is $\bar{x} < 27.28$.

(i) If $\mu = 31$,

$$\beta = 1 - \Phi\left(\frac{27.28 - 31}{1.658}\right)$$
$$= 0.9876.$$

(ii) If $\mu = 29$,

$$\beta = 1 - \Phi\left(\frac{27.28 - 29}{1.658}\right)$$
$$= 0.8502.$$

The full operating characteristic curves for both upper- and lower-tail tests, and for the two-tailed tests, are given in Fig. 64.

Notice that if the true value of μ is greater than the hypothesised value (i.e., 30) then the upper-tail test is more *powerful* (i.e., has a smaller β) than the two-tail test. In this situation the lower-tail test is almost useless, since β is nearly one. Clearly converse conclusions can be drawn when $\mu < 30$ by examining the left-hand half of the diagram. As a matter of fact, for a given level of significance, a, it can be shown that a one-tailed test is always more powerful than any other. Fig. 64 illustrates that a one-tailed test, unlike the two-tailed test, is biased in the sense that the probability of accepting the null hypothesis, H_0, is not greatest when H_0 is actually true.

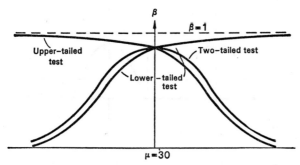

FIG. 64.—Operating characteristic curves for upper-, lower-, and two-tailed tests of H_0: $\mu = 30$, all with $a = 0.05$.

12. Control of Type I and Type II errors. It should be clear from the tests that we have conducted that the a or Type I error is under the tester's control. The Type II error in general is not. It is, therefore, of great help if the hypothesis can be so designed that a high (or unknown) value of β is unimportant. If, for example, the average daily takings in a supermarket must be £800 in order to break even, the Type II error in the test

$$H_0: \mu = 800$$
$$H_1: \mu > 800$$

is unimportant. On the other hand, the size of the Type II error in the test

$$H_0: \mu = 800$$
$$H_1: \mu < 800$$

is absolutely crucial, since incorrect acceptance of H_0 when in fact H_1 is true would give the management of the supermarket a false view of their financial position.

The three variables a, β, and n are interrelated, and for a fixed value of n a decrease in β can only be achieved at the expense of an increase in a (*see* Fig. 65 (*a*) and (*b*)).

However, an improvement is made if the sample size is increased. In the specific case of the sample mean this has the effect of reducing its variance (since increasing n decreases s^2/n), thereby enabling a reduction in β for the same level of a (*see* Fig. 65 (*c*)).

By way of illustration, our ubiquitous accountant (Section 1) has now come to the conclusion that the average time for debt-collection is 34 days. Unfortunately, a rival within the company claims that in fact the average is 37, and that our accountant ought to take steps to reduce it. Skilful statistician that he is, our accountant sets up the hypotheses

$$H_0: \mu = 34.$$
$$H_1: \mu = 37.$$

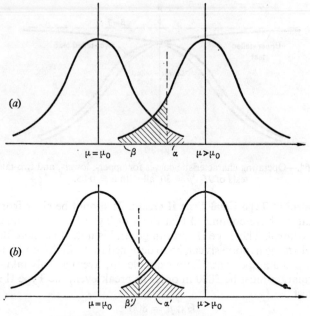

(a) and (b) show effect on a of decreasing β.

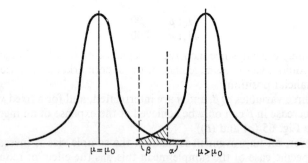

(c) Effect on a and β of increasing n.

Fig. 65.—Effect on a of decreasing β and effect on a and β of increasing n.

He decides on a significance level of $a = 0.02$, since he wants the probability of admitting himself wrong, when in fact he is correct, to be small. Nor, however, does he want a large probability of accepting H_0 when it is false, since this would lead to undesirable consequences in the future when his mistakes came to light. Accordingly the maximum Type II error he is prepared to accept is $\beta = 0.03$. Fortunately

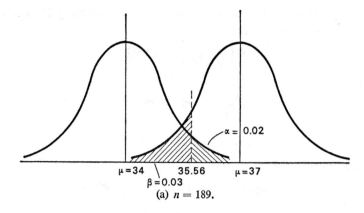

(a) $n = 189$.

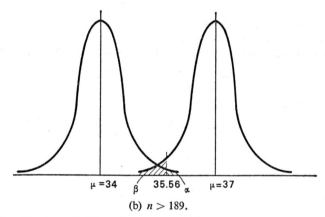

(b) $n > 189$.

FIG. 66.—Control of Type I and Type II errors by suitable choice of n.

there is agreement between his rival and himself that the population variance is 110.

With a sample size of n, and a sample mean of $\bar{x}$ his decision rule—bearing in mind that $a = 0.02$—is:

Reject H_0 if

$$\frac{(\bar{x} - 34)}{\sqrt{\left(\frac{110}{n}\right)}} \geqslant 2.05,$$

i.e., if

$$\bar{x} \geqslant 34 + 2.05\sqrt{\left(\frac{110}{n}\right)}. \tag{i}$$

The Type II error is then the probability of accepting H_0 when $\mu = 37$. Referring to Fig. 66,

$$\Phi\left(\frac{\bar{x} - 37}{\sqrt{(110/n)}}\right) \leqslant 0.03$$

i.e.,

$$\frac{\bar{x} - 37}{\sqrt{(110/n)}} \leqslant -1.88$$

or

$$\bar{x} \leqslant 37 - 1.88\sqrt{(110/n)}. \tag{ii}$$

Equating (*i*) and (*ii*),

$$34 + 2.05\sqrt{(110/n)} = 37 - 1.88\sqrt{(110/n)}$$
$$3 = 3.93\sqrt{(110/n)}$$
$$n = (3.93)^2 \times 110/9$$
$$= 188.8.$$

By taking a sample of size 189 he ensures that his Type I and Type II errors are both as small as he requires.

The critical value of $\bar{x}$ in his decision rule can now be seen to be:

$$\bar{x} = 34 + 2.05\sqrt{(110/189)}$$
$$= 35.56.$$

As a check,

Type I error
$$= 1 - \Phi\left\{\frac{35.56 - 34}{\sqrt{(110/189)}}\right\}$$
$$= 1 - \Phi(2.05)$$
$$= 0.02$$

and

Type II error
$$= P(\bar{x} < 35.56 \,|\, \mu = 37)$$
$$= \Phi\left\{\frac{35.56 - 37}{\sqrt{(110/189)}}\right\}$$
$$= \Phi(-1.88)$$
$$= 0.03$$

as required.

With the same critical value of 35.56, an increase of n beyond 189 will result in a decrease in both α and β (*see* Fig. 66).

CHAPTER 9

CONTINGENCY TABLES AND ANALYSIS
OF VARIANCE

INTRODUCTION

All our tests have been concerned with making inferences about a
population parameter, or with comparing two such parameters. None,
however, enables us to test hypotheses concerning several parameters.
The techniques to be developed in this chapter will correct this
omission.

Although *analysis of variance* (ANOVA) is a vital tool in scientific
work where, for example, the effects of different chemicals, drugs,
fertilisers or insecticides are to be compared, in the business world its
applications have not been widespread. Partly for this reason, and
partly because it is embedded in the larger subject of experimental
design and analysis, we shall not explore ANOVA in great detail.
References are provided in the Bibliography, p. 298 for the benefit of
the reader interested in pursuing the subject further.

CONTINGENCY TABLES

1. Dependence and independence. It is sometimes useful to be able to
detect whether two characteristics of a population are independent of
each other or not. As an insurance underwriter, for example, you
would be interested to know whether age of car driver, and severity of
accident (and therefore size of claim) were independent. Alternatively,
the accountant might look for the dependence or otherwise of the
debt-collection time and the size of discount offered.

To be sure that our understanding of independence is precise, let us
assume that the population is divided into those members with
attribute I, and those without. If the proportion of members of the
first category which additionally possess attribute II is the same as the
proportion of the second category which have attribute II, then the
attributes are independent (*cf.* Chapter 3, Section 2).

Consider Table 28, which concerns the 1200 current accounts at the
Natclay Bank Limited. This table is known as a *contingency* table.
The population of accounts is broken down by age of account-holder,
and by their record of overdrawals over the last 12 months.

The proportion of the customers of 25 or under who overdraw
frequently is $45/150 = 0.30$. The proportion with frequent over-

Table 28. Breakdown of accounts of Natclay Bank Ltd. (particular branch)

(A) Age of customer	Very rarely overdrawn	Sometimes overdrawn	Frequently overdrawn	Total
$A \leqslant 25$	35	70	45	150
$25 < A \leqslant 40$	105	210	135	450
$A > 40$	140	280	180	600
Total	280	560	360	1200

drawals is exactly the same in the second age group, and in the third age group, *i.e.* 135/450 = 0·30 and 180/600 = 0·30 respectively. Looking at the class of customer who is sometimes overdrawn, we find that the proportions in the three age categories are again all equal, *i.e.* 70/150 = 210/450 = 280/600 = 7/15. Finally, similar results appear in the "rarely overdrawn" class. We conclude that the factors "age of account-holder" and "frequency of overdrawal" are independent. Notice that analysing the table by rows (*i.e.*, taking each age category in turn, and calculating the proportions of it which lie in the various overdrawal categories) leads to the same conclusion.

2. Observed and expected frequencies. A less artificial situation than that represented in Table 28 is given in Table 29. Here there is not

Table 29. Breakdown of sample of Natclay Bank Ltd. head office accounts

(A) Age of customer	Very rarely overdrawn	Sometimes overdrawn	Frequently overdrawn	Total
$A \leqslant 25$	25	84	58	167
$25 < A \leqslant 40$	125	198	103	426
$A > 40$	115	303	189	607
Total	265	585	350	1200

such a clear relationship between the cell-frequencies. Furthermore, we shall assume that the 1200 accounts are a sample from the entire customer records held at the bank's head office.

Note that this data is derived from a sample, and that the various sampling proportions are therefore estimates of the corresponding proportions in the whole population of accounts. We are now interested in knowing whether there are any significant differences between sampling proportions which are unlikely to occur as a result of sampling error alone. The hypotheses to be tested are therefore:

H_0: "Age" and "frequency of overdrawal" are independent.
H_1: "Age" and "frequency of overdrawal" are not independent.

From a sample of 1200 customers 265 were rarely overdrawn. In the population, therefore, our estimate of the probability of an individual account falling into this category is $265/1200 = 0.2208$.

Furthermore, 167 customers in the sample were 25 or under.

Assuming H_0 to be true, we would expect that $167 \times 0.2208 = 36.87$ of the customers of 25 or under would be rarely overdrawn. The observed figure is 25. Similarly we would expect there to be

$$426 \times 265/1200 = 94.07 \text{ customers}$$

in the second age group who are rarely overdrawn. Proceeding in this manner throughout Table 29, we arrive at the expected cell-frequencies shown in Table 30.

Table 30. Expected frequencies in Natclay sample

(A) Age of customer	Very rarely overdrawn	Sometimes overdrawn	Frequently overdrawn	Total
$A \leqslant 25$	36·88	81·41	48·71	167
$25 < A \leqslant 40$	94·07	207·68	124·25	426
$A > 40$	134·05	295·91	177·04	607
Total	265	585	350	1200

The occurrence of fractional parts of customers need cause no alarm if we think of the figures as representing expected values over a series of possible samples of size 1200.

A few moments' thought will convince the reader that the manner in which the expected cell-frequencies are computed maintains the row and column totals at their previous values. For example, in the second column

$$81·41 = \frac{167 \times 585}{1200}$$

$$207·68 = \frac{426 \times 585}{1200}$$

$$295 \cdot 91 = \frac{607 \times 585}{1200}$$

$$81 \cdot 41 + 207 \cdot 68 + 295 \cdot 91 = (167 + 426 + 607) \times \frac{585}{1200}$$

$$= \frac{1200 \times 585}{1200} = 585.$$

The expected frequencies, calculated on the assumption that H_0 is true, differ from the observed (actual) frequencies. Are the differences sufficient to enable us to reject H_0?

3. Chi-squared statistic. We introduce at this point a new statistic known as the χ^2 (chi-squared). Denoting the observed and expected frequencies in the ith and jth column by O_{ij}, E_{ij} respectively we calculate

$$\chi^2 = \frac{(O_{ij} - E_{ij})^2}{E_{ij}}.$$

This rather curious quantity can be explained by noting that $(O_{ij} - E_{ij})^2$ is a measure of the deviation of observed from expected values in cell i, j. This is squared to remove any negative sign, since subsequent summation would otherwise cause cancellation. Because squaring results in a disproportionate weight being attached to the larger deviations we divide by E_{ij}. The total "deviation from H_0" is then found by summing terms over all cells.

Clearly, the larger the χ^2 value, the greater the overall deviation of observed values, and the better the case for rejecting H_0. As yet, however, we have no way of deciding how large a χ^2 value must be in order to indicate significant deviation from the null hypothesis.

It can be shown that, if H_0 is true, the χ^2-statistic has a sampling distribution of a particular type. Since the χ^2-statistic can assume any positive value, its distribution has the same property. The exact form of the distribution is dependent once more on the concept of degrees of freedom, the number of which can be found by considering the constraints imposed on the sample information in the process of calculating the χ^2-statistic.

Suppose that the contingency table contains r rows and c columns. In calculating the expected cell-frequencies for a particular row one constraint is imposed (i.e., that their total has the value of the sample row total). Since similar comments apply to each column, the row and column totals provide altogether $r + c$ constraints. The grand total (i.e., sample size) provides a constraint, which is automatically satisfied if either all the row constraints or all the column constraints are satisfied. It has, therefore, already been included in r and in c, and since we do not want to count it twice, we arrive at a figure of $r + c - 1$ constraints on the rc cell-values. The number of degrees of freedom is therefore:

$$rc - r - c + 1 = r(c - 1) - (c - 1)$$
$$= (r - 1)(c - 1).$$

The sampling distribution of the χ^2-statistic, which is referred to as the χ^2-*distribution* is shown in Fig. 67 for various numbers of degrees of freedom.

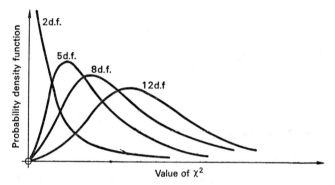

FIG. 67.—χ^2-distribution for various degrees of freedom.

The values of the χ^2-statistic and its associated probability distribution function are given in Appendix V for different degrees of freedom, n. The expected (mean) value of the χ^2-distribution is n, and the variance $2n$.

4. Significance tests using the χ^2-statistic. Knowledge of the sampling distribution of the χ^2-statistic when H_0 is true enables us to conduct significance tests in the now familiar manner.

Returning to the illustration of Section 2 above:

(i) H_0: "Age" and "frequency of overdrawal" are independent.
   H_1: "Age" and "frequency of overdrawal" are not
                                        independent.

(ii) $a = 0.05$.

(iii) Test statistic: $\chi^2 = \dfrac{(O_{ij} - E_{ij})^2}{E_{ij}} \sim \chi^2 (r - 1)(c - 1)$

if H_0 is true. Now $r = 3$, $c = 3$, so there are

$$(3 - 1)(3 - 1) = 4 \text{ degrees of freedom.}$$

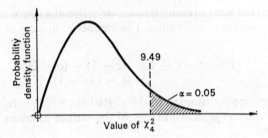

FIG. 68.—χ^2-distribution, four degrees of freedom.

(*iv*) Critical value of χ^2 is 9·49 (*see* Fig. 68).

(*v*) Sample results:

$$\chi^2 = \frac{(25 - 36 \cdot 88)^2}{36 \cdot 88} + \ldots + \frac{(189 - 177 \cdot 04)^2}{177 \cdot 04}$$
$$= 3 \cdot 8268 + 0 \cdot 0824 + 1 \cdot 7718 + 10 \cdot 1697 + 0 \cdot 4512$$
$$+ 3 \cdot 6343 + 2 \cdot 7072 + 0 \cdot 1699 + 0 \cdot 8079$$
$$= 23 \cdot 6212.$$

(*vi*) H_0 is rejected.

The detailed calculation of χ^2 using the formula has been shown in this example. However, for computational purposes the form,

$$\chi^2 = \Sigma\left(\frac{O_{ij}^2}{E_{ij}}\right) - n$$

where n is the sample size, is more convenient.

Note that we have used an upper-tailed test, since a large value of χ^2 is indicative of considerable difference between observed and expected values, and therefore of departure from the null hypothesis. However, the mean value of the sampling distribution of χ^2 is, in this example, 3. Accordingly, even if the null hypothesis is true, actually to obtain a value of χ^2 near zero is very unlikely. For example, the probability that χ_4^2 is less than 0·71 is only 0·05. Since small values of the χ^2-statistic are the result of little difference between observed and expected values, we seem to have a paradox. One might have thought that the smaller the sample χ^2-statistic (and therefore the better the match between observed and expected values) the greater the weight of evidence in favour of H_0.

The explanation lies in the fact that even when H_0 is true, one would expect sampling error to cause some deviation between observed and expected frequencies. A very small χ^2-statistic implies a suspiciously high degree of agreement. It may be, for example, that the sampling is faulty or that the data are incorrectly processed.

5. Test for equality of several proportions. We have already considered a test for the equality of two proportions (*see* Chapter 8, Section 8). The χ^2-statistic and its distribution enable us to test also for differences between several population proportions.

Suppose a sample of accounts was taken from each of four branches of Natclays Bank Limited, and that the proportion of accounts overdrawn at the time of sampling found in each case (Table 31).

Table 31. Observed breakdown of accounts at four Natclay branches

Branch	1	2	3	4	Total
No. of accounts overdrawn	25	17	31	32	105
No. of accounts not overdrawn	76	70	87	60	293
Total	101	87	118	92	398
Proportion	0·2475 ($=p_1$)	0·1954 ($=p_2$)	0·2627 ($=p_3$)	0·3478 ($=p_4$)	

Our null hypothesis is that the proportions of overdrawn accounts in the four branches are all equal to p, say, the best available estimate of which is $105/398 = 0.2638$. Using this estimate, we would expect there to be $101 \times 0.2638 = 26.65$ overdrawn accounts in the sample of 101 taken from branch 1. Proceeding in this way, we arrive at the expectations shown in Table 32.

Table 32. Expected breakdown of accounts at four Natclay branches

Branch	1	2	3	4	Total
Expected no. of accounts overdrawn	26·65	22·95	31·13	24·27	105
Expected no. of accounts not overdrawn	74·35	64·05	86·87	67·73	293
Total	101	87	118	92	398

The test can now be conducted:

(i) $\qquad H_0: p_1 = p_2 = p_3 = p_4 \ (=p)$.
$\qquad\qquad H_1:$ At least one pair of proportions unequal.

(ii) $\qquad a = 0.02$.

(*iii*) Test statistic:

$$\chi^2 = \sum \frac{(O_{ij} - E_{ij})^2}{E_{ij}} \sim \chi^2(r-1)(c-1) = \chi^2(4-1)(2-1)$$
$$= \chi_3^2.$$

(*iv*) Critical value of χ^2-statistic is 9·837.

(*v*) Sample results: $\chi^2 = \dfrac{(26 \cdot 65 - 25)^2}{26 \cdot 65} + \ldots + \dfrac{(67 \cdot 73 - 60)^2}{67 \cdot 73}$

$$= 5 \cdot 58.$$

(*vi*) H_0 is accepted.

6. The continuity correction. In the methods of both Sections 2 and 5 above a correction needs to be applied to the computed χ^2-statistic if either the frequency table is only 2×2, or the sample size is so small that at least one of the expected cell-frequencies is 5 or less. The reason for this rests on the fact that the χ^2-statistic has only an approximately χ^2-distribution, and when the table or sample is small this has to be duly recognised.

The correction involves replacement of the previous χ^2-statistic by:

$$\chi^2 = \frac{\{ |(O_{ij} - E_{ij})| - \frac{1}{2}\}^2}{E_{ij}}.$$

Here the modulus sign | | signifies the *positive* difference between O_{ij} and E_{ij}. The procedure is illustrated in the subsequent worked example.

Worked Example

A survey is carried out on a random sample of 44 students who graduated from Ardour University with bachelors' degrees in accounting six years ago. The purpose of the survey is to ascertain whether there is any relationship between degree classification and future income. The results of the survey are given in Table 33.

Table 33. Breakdown of incomes of Ardour University graduates

Degree	Low income	Medium income	High income	Total
First Class Hons.	7	2	3	12
Second Class Hons.	7	4	3	14
Third Class Hons.	2	7	9	18
Total	16	13	15	44

Answer

The expected frequencies are shown in Table 34.

Table 34. Expected breakdown of incomes of Ardour University graduates

Degree	Low income	Medium income	High income	Total
First Class Hons.	4·36	3·54	4·09	12
Second Class Hons.	5·09	4·14	4·77	14
Third Class Hons.	6·55	5·32	6·14	18
Total	16	13	15	44

(*i*) H_0: Income and degree classification independent.

H_1: Income and degree classification not independent.

(*ii*) $a = 0·10$.

(*iii*) Test statistic

$$\chi^2 = \sum \frac{\{|(O_{ij} - E_{ij})| - \frac{1}{2}\}^2}{E_{ij}} \sim \chi^2_{(3-1)(3-1)} = \chi_4^2$$

The correction is used since the table contains several cell-frequencies of 5 or less.

(*iv*) Critical value of χ^2-statistic is 7·78.

(*v*) Sample results:

$$\chi^2 = \frac{(2·64 - 0·5)^2}{4·36} + \frac{(1·54 - 0·5)^2}{3·54} + \frac{(1·09 - 0·5)^2}{4·09}$$

$$+ \frac{(1·91 - 0·5)^2}{5·09} + \frac{(0·14 - 0·5)^2}{4·14} + \frac{(1·77 - 0·5)^2}{4·77}$$

$$+ \frac{(4·55 - 0·5)^2}{6·55} + \frac{(1·68 - 0·5)^2}{5·32} + \frac{(2·86 - 0·5)^2}{6·14}$$

$$= 1·0504 + 0·3055 + 0·0851 + 0·3906 + 0·0313$$
$$+ 0·3381 + 2·5042 + 0·2617 + 0·9071$$

$$= 5·8740.$$

(*vi*) H_0 is accepted.

Note, however, that if the continuity correction had not been used, the χ^2-statistic would have been 8·96, and the incorrect decision to reject H_0 would have been made.

Exercises on Section 6

1. The following table gives the number of students who passed and failed a particular subject in (*i*) Course work, (*ii*) Examination. Use a χ^2

significance test to determine whether there is any association between the results in course work and examinations.

		Course Work		
		Pass	Fail	Total
	Pass	610	140	750
Exam	Fail	190	60	250
	Total	800	200	1000

2. A bureau of business research polled a sample of 500 men engaged in different fields of business, to determine whether there were any differences in attitudes towards the prospects of overall business activity in the coming year. The results of this poll were as follows:

	Bankers	Manufacturers	Retailers	Farmers
Increased activity	40	55	80	75
Decreased activity	15	30	40	50
No appreciable change	20	25	30	40

Does there appear to be a significant difference of opinions between the four groups?

3. A student, looking for an easy statistics teacher, was told by a departmental office that all three statistics teachers passed the same proportion of students. The student did some research and came up with the following results:

Student performance	Teacher A	Teacher B	Teacher C	Total
Number passed	42	43	38	123
Number failed	8	5	14	27
Total	50	48	52	150

Should the student believe what the office told him? Use a 0·05 level of significance.

4. The credit manager of Spendthrift Supermarkets has been examining the accounts receivable for the five stores in Ambridge. He notes that the

proportions of accounts six or more months' delinquent seem to vary. On the basis of the following information, does the rate of delinquency of accounts receivable vary between stores in Ambridge?

Store	Number of Accounts		
	Delinquent at least 6 months	Not delinquent 6 months	Total
A	25	230	255
B	10	179	189
C	14	214	228
D	9	158	167
E	9	109	118
Total	67	890	957

5. An analysis of the costs of idle machine time in three of Rotrite Incorporated's factories reveals the following:

Plant	Breakdown	Cause of Idle Time		
		Operator error	Lack of materials	Reloading machine
A	9·2	13·5	8·3	5·9
B	9·6	11·5	7·8	5·1
C	6·6	9·6	9·5	3·2

Is there any reason to suppose that cause of idle time varies between plants?

6. A subscription service stated that preferences for different national magazines were independent of geographical location. Data on the choice of favourite out of three magazines were as follows:

Region	Magazine X	Magazine Y	Magazine Z	Total
Lancashire	75	50	175	300
Warwickshire	120	85	95	300
Devon	105	110	85	300
Total	300	245	355	900

Would you agree with the subscription service's assertion (at the 0·05 significance level)?

ANALYSIS OF VARIANCE

7. One-way analysis of variance. To test for differences between the means of several populations requires a different approach from that used in testing for differences between proportions.

Let us take the case of 18 machines belonging to an engineering company. The company employs three maintenance engineers, each of whom has responsibility for six machines. Records have been kept over the last year of the maintenance costs of the eighteen machines (*see* Table 35).

Table 35. Maintenance costs per year (£)

| | Engineer | | |
	I	*II*	*III*
	91	96	97
	94	91	99
	97	98	96
	91	92	95
	92	98	94
	95	95	99
Mean	93·33	95·00	96·67

The company's accountant is interested in knowing if there is a significant difference in maintenance costs between the groups of machines (*i.e.*, between engineers).

We may treat the six costs attributable to an engineer as a random sample from the population of costs for which he is responsible. If μ_1, μ_2, μ_3 are the respective population means, our hypotheses are:

$$H_0: \mu_1 = \mu_2 = \mu_3 = \mu, \text{ say.}$$
$$H_1: \mu_1, \mu_2, \mu_3 \text{ are not all equal.}$$

To test these hypotheses we analyse the variance between the eighteen data values. First, there exists a variance *within* each of the three sets of figures, attributable simply to random effects. Second, there is a variance *between* the groups which is attributable to genuine differences between μ_1, μ_2, μ_3.

In the next two sections we demonstrate how the two sources of variance may be isolated and measured.

8. Variance within groups. Taking group I first of all, the estimate of the variance of population I is:

$$\frac{(91 - 93·33)^2 + \ldots + (95 - 93·33)^2}{(6 - 1)}.$$

Similarly, an estimate of the variance of population II is provided by:

$$\frac{(96 - 95 \cdot 00)^2 + \ldots + (95 - 95 \cdot 00)^2}{(6 - 1)}$$

and of population III by:

$$\frac{(97 - 96 \cdot 67)^2 + \ldots + (99 - 96 \cdot 67)^2}{(6 - 1)}.$$

Since the three quantities above each estimate the same parameter (*i.e.*, the variance in costs due to random error alone) we may pool them (*cf.* Chapter 7, Section 13) to form the single estimate:

$$\frac{(91 - 93 \cdot 33)^2 + \ldots + (96 - 95 \cdot 00)^2 + \ldots + (97 - 96 \cdot 67)^2 + \ldots}{(6 - 1) + (6 - 1) + (6 - 1)}$$

$$= 6 \cdot 31.$$

9. Variance between groups. To identify the variance *between* groups, let us eliminate the *within* groups variance by replacing each cost by the group average (*see* Table 36).

Table 36. Modified maintenance costs—eliminating within groups variation (conceptual)

	Engineer		
	I	*II*	*III*
	93·33	95·00	96·67
	93·33	95·00	96·67
	93·33	95·00	96·67
	93·33	95·00	96·67
	93·33	95·00	96·67
	93·33	95·00	96·67
Mean	93·33	95·00	96·67

On the assumption that H_0 is true, *i.e.* $\mu_1 = \mu_2 = \mu_3 = \mu$ the differences between groups are attributable to random error alone, and the three groups are then independent samples from the same population having mean μ. The variance between the sample *means*

$$= \frac{(93 \cdot 33 - 95)^2 + (95 \cdot 00 - 95)^2 + (96 \cdot 67 - 95)^2}{2}$$

$$= 2 \cdot 789.$$

Since each sample mean results from a sample size of six, the estimate of population variance using differences between groups is:

$$6\left\{\frac{(93 \cdot 33 - 95)^2 + (95 \cdot 00 - 95)^2 + (96 \cdot 67 - 95)^2}{2}\right\} = 16 \cdot 73$$

(cf. var $\bar{x} = \sigma^2/n$, $\therefore$ var $x = \sigma^2$).

This estimate is clearly much larger than the previous one.

We need to know, however, whether it is sufficiently large for us to be able to conclude that the variance between groups must be attributable not only to random error, but also to the fact that at least one of the samples contains values which are, on the whole, significantly greater or less than those of the other samples.

10. The sampling distribution of the variance ratio (the F-distribution).

Let us look more closely at the two quantities calculated in Sections 8 and 9 above, both of which, under H_0, estimate the (assumed equal) variance of the populations from which the samples are drawn. The first estimate has 15 degrees of freedom, since the 18 data values are subject to three constraints in the form of sample means. Alternatively, we may regard three degrees of freedom as having been lost in estimating the respective population means.

The second estimate has two degrees of freedom, since the three quantities used in its calculation (i.e., sample means) are subject to the constraint that their mean value is 95·00. Alternatively, one of the three degrees of freedom is sacrificed to estimate the overall mean.

Using the two estimates of the common population variance we now calculate the so-called F-statistic:

$$F = \frac{\text{estimate of population variance using between sample differences}}{\text{estimate of population variance using within sample differences}}$$

$$= \frac{16 \cdot 73}{6 \cdot 31}$$

$$= 2 \cdot 65.$$

The F-statistic, or variance ratio, formed from the ratio of two independent estimates of the same variance, has a sampling distribution known as an F-distribution (cf. χ^2-statistic and χ^2-distribution). The F-distributions form a family, with a particular member of the family F_{v_1, v_2} being uniquely defined by two parameters, viz. the numbers of degrees of freedom v_1, v_2 of the estimates in the numerator and denominator of the ratio respectively. Fig. 69 gives the shape of the F-distribution for a few values of these parameters.

As we should expect, the mean value of an F-distribution is in the region of unity, since on average the ratio of two independent estimates of the same variance will, because of sampling error, sometimes exceed one and sometimes be less than one. In fact, the mean value is exactly $v_2/(v_2 - 2)$.

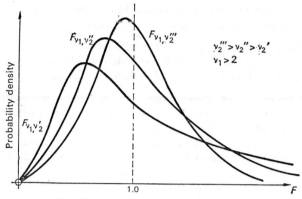

Fig. 69.—General form of F-distribution.

In our particular example, we have

$$F_{2, 15} = 2 \cdot 65.$$

The upper 10% and 5% points of this distribution (*see* Fig. 70) are $2 \cdot 70$ and $3 \cdot 68$. On the assumption that H_0 is true, the occurrence of a variance ratio as large as $2 \cdot 65$ is not sufficiently unlikely for us to be able to reject H_0 either at the 5% or the 10% levels of significance.

Notice that in calculating the variance ratio the larger estimate appeared in the numerator, giving a ratio exceeding one. This is normal practice, since tables of the F-distribution usually give upper-tail areas (suitable for an upper-tail test). It is important, however, to ensure that the correct F-distribution tables are consulted (*see* Appendixes VI and VII). In this case, since the estimate with 2 degrees of freedom appeared in the numerator, the appropriate tables are of $F_{2, 15}$, and not $F_{15, 2}$.

11. Sums of squares. The sums of the squared deviations from a mean value, or *sums of squares* provide some help in isolating the sources of

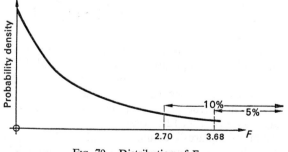

Fig. 70.—Distribution of $F_{2, 15}$.

variance amongst a set of data values. In Section 8, for example, the sums of squares within each group are respectively,

(i) $(91 - 93 \cdot 33)^2 + \ldots + (95 - 93 \cdot 33)^2 = 29 \cdot 33.$
(ii) $(96 - 95 \cdot 00)^2 + \ldots + (95 - 95 \cdot 00)^2 = 44 \cdot 00.$
(iii) $(97 - 96 \cdot 67)^2 + \ldots + (99 - 96 \cdot 67)^2 = 21 \cdot 33.$

Their total, i.e. the *within groups sum of squares* is the quantity which is divided by the number of degrees of freedom within groups (i.e., 15) to obtain an estimate of the population variance.

Similarly, in Section 9 the *between groups sum of squares* is divided by the number of degrees of freedom between groups (i.e., 2) to obtain a second estimate of the population variance.

Finally, the *total sum of squares*, defined by sum of the squared deviations of each data value from the grand mean, has a value

$$(91 - 95)^2 + \ldots + (99 - 95)^2 = 128 \cdot 00$$

and can be divided by the total number of degrees of freedom (i.e., $18 - 1 = 17$, since the data values are subject to one constraint, or used to estimate one parameter, viz. the mean) to provide a third estimate of population variance.

These results are summarised in an *ANOVA table* (Table 37).

Table 37. Analysis of variance table for maintenance costs

Source of variation	Sum of squares	Degrees of freedom	Mean square (i.e. variance estimate)	F (variance) ratio
Between groups	$\Sigma\left(\dfrac{T_i^2}{n_i}\right) - \dfrac{T^2}{N} = 33 \cdot 34$	2	$16 \cdot 67$	$\dfrac{16 \cdot 67}{6 \cdot 31} = 2 \cdot 64$
Within groups	$S - \Sigma\left(\dfrac{T_i^2}{n_i}\right) = 94 \cdot 66$	15	$6 \cdot 31$	
Total	$S - \dfrac{T^2}{N} = 128 \cdot 00$	17	$7 \cdot 53$	

The fact that the total sum of squares is equal to the sum of the within groups and between groups sums of squares, and that the same property is possessed by the corresponding numbers of degrees of freedom, is no accident. The fact that these relationships can be shown to always hold true provides a very useful tool in more advanced analysis (*see* Section 12 below).

In the "sum of squares" column of Table 37, some symbolic representations have been given. These indicate more efficient ways of calculating the three quantities than those given in the illustration, and have the following meanings:

T_i = group sum.
n_i = group size.
T = grand total of all measurements.
N = total sample size.
S = sum of squares of all measurements.

The worked example below illustrates the entire procedure.

Worked Example

Miss Smith, supervisor of a secretarial pool, has four typists under her supervision. She is concerned with the length of their coffee break, so she times them. Her observations for each girl, recorded in minutes, are given below. Can the differences in average time that the four typists spent on coffee breaks be explained by chance variation?

	Typist		
I ($n_1 = 9$)	II ($n_2 = 12$)	III ($n_3 = 11$)	IV ($n_4 = 7$)
27	20	24	20
35	22	22	18
18	30	25	26
24	27	25	19
28	22	20	26
32	24	21	24
16	28	34	26
18	21	18	—
25	23	32	159
—	18	23	—
223	30	22	
—	32	—	
	—	266	
	297	—	
	—		

NOTE: This example involves four samples of different sizes, unlike the previous case. The modifications to the previous example are natural ones. Thus the (pooled) estimator of population variance from *within* samples is:

$$s^2 = \frac{(n_1 - 1)s_1{}^2 + (n_2 - 1)s_2{}^2 + (n_3 - 1)s_3{}^2 + (n_4 - 1)s_4{}^2}{(n_1 - 1) + (n_2 - 1) + (n_3 - 1) + (n_4 - 1)}$$

and the estimator from between samples is:

$$\frac{n_1(\bar{x} - \bar{\bar{x}})^2 + n_2(\bar{x}_2 - \bar{\bar{x}})^2 + n_3(\bar{x}_3 - \bar{\bar{x}})^2 + n_4(\bar{x}_4 - \bar{\bar{x}})^2}{(4 - 1)}$$

where $\bar{\bar{x}}$ is overall mean.

Answer
Between groups (i.e., typists)

Sum of squares

$$= \left[\frac{T_1^2}{n_1} + \frac{T_2^2}{n_2} + \frac{T_3^2}{n_3} + \frac{T_4^2}{n_4} \right] - \frac{T^2}{N}$$

$$= \frac{(223)^2}{9} + \frac{(297)^2}{12} + \frac{(266)^2}{11} + \frac{(159)^2}{7} - \frac{(945)^2}{39}$$

$$= 5525 \cdot 4 + 7350 \cdot 8 + 6432 \cdot 4 + 3611 \cdot 6 - 22,898 \cdot 1$$
$$= 22 \cdot 1.$$

Degrees of freedom $= 4 - 1 = 3$.

Mean Square $= 7 \cdot 37$.

Within Groups

Sum of squares

$$= S - \sum \frac{T_i^2}{n_i}$$
$$= 23,799 - 22,920 \cdot 2$$
$$= 878 \cdot 8.$$

Degrees of freedom $= 39 - 4 = 35$.

Mean Square $= 25 \cdot 11$.

The ANOVA results are shown in Table 38.

Table 38. ANOVA table for typists' coffee break times

Source of variation	Sums of squares	Degrees of freedom	Mean square	F (variance) ratio
Within groups	878·8	35	25·11	$\frac{25 \cdot 11}{7 \cdot 37} = 3 \cdot 40$
Between groups	22·1	3	7·37	
Total	900·9	38		

Test

(i) $H_0 \colon \mu_1 = \mu_2 = \mu_3 = \mu_4$.
              $H_1 \colon$ at least one pair of $\mu_1, \mu_2, \mu_3, \mu_4$ unequal.

(ii) $a = 0 \cdot 05$.

(iii) Test statistic: F-statistic $\sim F_{35, \, 3}$.
                          Critical value: $8 \cdot 60$.
                          Reject H_0 if $F > 8 \cdot 60$.

(iv) Sample results: $F = 3 \cdot 40$.

(v) H_0 is not rejected.

Exercises on Section 11

7. A manufacturer has a choice of three sub-contractors from whom to buy parts. The manufacturer, before deciding from whom he will buy, purchases five batches from each sub-contractor. There are the same number in each batch. The number of defectives per batch is given in the following table:

Sub-contractor		
A	*B*	*C*
35	15	25
25	20	40
30	25	40
35	15	35
20	30	30

Would you conclude that there is no real difference among these sub-contractors in the average number of defectives produced per batch? Use a 0·05 level of significance.

8. The following are three different years' returns on capital of five industrial groups.

Food manufacturers	Retail outlets	Light engineering	Heavy engineering	Brewing
12·6	14·8	20·4	11·9	15·6
15·2	17·1	18·2	12·5	18·4
13·9	17·6	17·2	15·5	16·1

Can the differences between these returns be attributed to chance?

9. The table below gives the monthly sales of four salesmen.

Salesman 1	Salesman 2	Salesman 3	Salesman 4
2	3	6	5
3	4	8	5
1	3	7	5
3	5	4	3
1	0	10	2

Is this sufficient evidence that there is a significant difference between the salesmen?

10. A trial is carried out to assess the effects on fuel consumption of two experimental petrol additives. Five observations are made of miles per gallon, obtained using each of two additives, *A* and *B*, and no additive, *C* with the following results

Treatment	Mean	Variance
A	28·9	1·1
B	28·5	1·4
C	27·5	1·1

Draw up an analysis of variance table and, assuming the observations are normally distributed with constant variance, test for differences between treatment means.

11. As head of a department of a consumers' research organisation, you have the responsibility for testing and comparing lifetimes of light bulbs for four brands of bulbs. Suppose you test the lifetime of three bulbs of each of the four brands. Your test data are as follows, each entry representing the lifetime of a bulb, measured in hundreds of hours.

Brand

A	B	C	D
20	25	24	23
19	23	20	20
21	21	22	20

Can we assume that the mean lifetimes of the four brands are equal?

12. Two-way ANOVA. In our earlier examples we were interested in testing for the significance of differences in one factor only, *e.g.* performance of engineers, length of typists' coffee breaks, etc. Now let us look at a more complex situation where two factors are present.

Returning to our maintenance costs (Table 35) of Section 7 above, let us suppose that each group contains six different types of machine, *A, B, C, D, E,* and *F*. For convenience, the figures are reproduced in Table 39.

Table 39. Maintenance costs: by machine types and by engineers

Machine type	Engineer I	Engineer II	Engineer III	Total
A	91	96	97	284
B	94	91	99	284
C	97	98	96	291
D	91	92	95	278
E	92	98	94	284
F	95	95	99	289
Total	560	570	580	1710

The quantity previously calculated as the within groups sum of squares is now seen to include also any variance which is due to differences between machine types. In order to separate out this possible component of the variance, we calculate a between types sum of squares in exactly the same manner as we previously calculated the between groups sum of squares:

Between Types

Sum of squares
$$= \frac{(284)^2}{3} + \frac{(284)^2}{3} + \frac{(291)^2}{3} + \frac{(278)^2}{3}$$
$$+ \frac{(284)^2}{3} + \frac{(289)^2}{3} - \frac{(1710)^2}{18}$$
$$= 26{,}885{\cdot}3 + 26{,}885{\cdot}3 + 28{,}227 + 25{,}761{\cdot}3$$
$$\qquad + 26{,}885{\cdot}3 + 27{,}840{\cdot}3 - 162{,}450$$
$$= 34{\cdot}5.$$

Degrees of freedom $= 6 - 1 = 5$.
Mean square $\qquad = 34{\cdot}5/5 = 6{\cdot}9$.

Between Groups (i.e., engineers)

Sum of squares
$$= \frac{(560)^2}{6} + \frac{(570)^2}{6} + \frac{(580)^2}{6} - \frac{(1710)^2}{18}$$
$$= 52{,}266{\cdot}7 + 54{,}150 + 56{,}066{\cdot}7 - 162{,}450$$
$$= 33{\cdot}4.$$

Degrees of freedom $= 3 - 1 = 2$.
Mean square $\qquad = 16{\cdot}7$.

Total

Sum of squares
$$= 91^2 + \ldots + 99^2 - \frac{(1710)^2}{18}$$
$$= 578 - \frac{(90)^2}{18}$$
$$= 128.$$

Degrees of freedom $= 17$.
Mean square $\qquad = 7{\cdot}53$.

Residual Sum of Squares

The additional effect of these calculations has been to divide the within groups sum of squares (all of which was previously attributed to random differences in the populations) into two components. The first component can now be explained by differences between types (sum of squares $= 34{\cdot}2$, degrees of freedom $= 5$). The second component (sum of squares $94{\cdot}6 - 34{\cdot}5 = 60{\cdot}1$, degrees of freedom

15 — 5 = 10) is known as the *residual*, and in the absence of other sources of variation is attributed to random variations in the population. It is therefore with this residual that the between groups and between types mean squares are compared for significance.

Table 40. Two-way ANOVA table of maintenance costs

Source of variation	Sums of squares	Degrees of freedom	Mean square	Variance ratio
Between groups	33·4	2	16·7	$\dfrac{16\cdot7}{6\cdot01} = 2\cdot78$
Between types	34·5	5	6·90	$\dfrac{6\cdot90}{6\cdot01} = 1\cdot15$
Residual	60·1	10	6·01	—
Total	128	17	7·53	—

The analysis is summarised in Table 40. Comparison of the two variance ratios with the upper 5% points of $F_{2, 10}$ and $F_{5, 10}$ (*i.e.*, 4·10 and 3·33) respectively, enables us to conclude that at this level of significance there are neither differences between engineers nor types of machine.

A second example is worked below for further clarification.

Worked Example

Records are kept on the sales of four salesmen for one week. Their sales for each day are shown below. Test at $a = 0\cdot01$ to see if there is a significant difference between salesmen, or between days of the week.

| Day of week | Salesman | | | |
	Jones	Smith	Walters	Brown
Monday	750	890	780	500
Tuesday	800	950	720	1000
Wednesday	850	640	500	800
Thursday	800	760	420	750
Friday	820	200	710	700

Answer

To simplify the calculation the data are coded by reducing all figures by a factor of 100. Since we shall eventually calculate variance *ratios* this will make no difference to the final results (*see* Table 41).

Table 41. Coded data for salesmen

Day	Salesman				Total
	Jones	Smith	Walters	Brown	
Mon.	7·5	8·9	7·8	5·0	29·2
Tues.	8·0	9·5	7·2	10·0	34·7
Wed.	8·5	6·4	5·0	8·0	27·9
Thurs.	8·0	7·6	4·2	7·5	27·3
Fri.	8·2	2·0	7·1	7·0	24·3
Total	40·2	34·4	31·3	37·5	143·4

Between salesmen sum of squares

$$= \frac{(40\cdot2)^2}{5} + \ldots + \frac{(37\cdot5)^2}{5} - \frac{(143\cdot4)^2}{20} = 8\cdot89.$$

Between days sum of squares

$$= \frac{(29\cdot2)^2}{4} + \ldots + \frac{(24\cdot3)^2}{4} - \frac{(143\cdot4)^2}{20} = 14\cdot55.$$

Total sum of squares

$$= 1095\cdot90 - 1028\cdot18 = 67\cdot72.$$

Residual sum of squares

$$= 67\cdot72 - 14\cdot55 - 8\cdot89 = 44\cdot28.$$

The analysis is completed in Table 42. The upper 5% points of $F_{3,\,12}$ and $F_{4,\,12}$ are 3·49 and 3·26. Comparing these with the variance ratios of 0·80 and 0·99 we see that there is no significant difference between salesmen or days.

Table 42. ANOVA table for data on salesmen

Source of variation	Sums of squares	Degrees of freedom	Mean square	Variance ratio
Between salesmen	8·89	3	2·96	$\frac{2\cdot96}{3\cdot69} = 0\cdot80$
Between days	14·55	4	3·64	$\frac{3\cdot64}{3\cdot69} = 0\cdot99$
Residual	44·28	12	3·69	
Total	67·72	19		

Exercises on Section 12

12. The Bigtown Transit Company is studying the fares distributions on different transportation routes at different times of year. Average fares (in pence) are shown below.

| Month of the year | Route number | | | | | |
	1	2	3	4	5	6
January	8·0	10·0	7·3	7·3	9·7	15·3
April	7·0	6·7	5·0	6·7	7·0	9·0
July	10·6	12·0	11·3	9·3	11·3	10·0
October	11·6	15·3	13·3	15·0	13·0	15·6

The company wants to know if the average fares vary by route and/or by time of year.

13. The average daily takings at five supermarkets over the last four weeks have been:

| Week | Supermarket | | | | |
	Cashco	Kwikcash	Spur	Waypole	Farewell
1	514	523	517	502	495
2	524	530	494	511	522
3	515	514	526	493	507
4	505	492	537	509	490

Do daily takings vary significantly between supermarkets and/or between weeks?

14. Four different machines are available for processing gudgeons. Each gudgeon has several operations performed on it on any one of the machines, and there are four orders in which these operations can be carried out. The number of defective gudgeons per batch of 100 are recorded:

| Order of processing | Machine | | | |
	A	B	C	D
I	2·36	2·44	2·60	2·67
II	3·02	3·01	3·16	3·39
III	3·16	3·20	3·34	3·47
IV	2·96	3·09	3·21	3·16

Is there any reason to suppose that the order of performing operations and/or the machine used affects percentage of defective gudgeons?

CHAPTER 10

NON-PARAMETRIC STATISTICS

INTRODUCTION AND DEFINITION

The methods of estimation and inference which we have discussed have often involved assumptions about the properties of the population being sampled. In addition, our work has centred on estimating the values of population parameters, or on testing hypotheses concerning those values.

Because of the restrictive nature of, for example, the assumption of near-normality necessary in small sample theory, alternative procedures have been developed based on *non-parametric statistics*. Statistics of this type are derived in the absence of assumptions about the shape or parameters of the population distribution.

These statistics, which also have the merit of computational simplicity, can be applied quite generally, and are particularly useful when exact numerical data is unavailable.

TESTS INVOLVING ONE SAMPLE

1. Runs test of randomness. We have spoken earlier of random number tables and the selection of random samples. There are two simple non-parametric statistics which enable us to test whether a series of numbers are in fact random.

The runs test is based on a sample of n observations recorded in the order in which they are made. Taking each pair of consecutive observations in turn, a plus or minus sign is inserted between them depending on whether the earlier observations is less than the later, or vice-versa. Defining a *run* as a continuous sequence of plus or minus signs, the number of runs, R, can be shown to have an approximately normal distribution with mean $\frac{1}{3}(2n - 1)$ and variance $\frac{1}{90}(16n - 29)$ if the sample is, in fact, random. This approximation is sufficiently close for most purposes when $n \geqslant 20$, and improves as n increases.

Consider the following daily sales (in £) recorded by a high-pressure salesman over a period of 25 days:

210	180	170	240	150	215	198	181	237
209	165	176	224	201	181	252	219	154
	197	235	182	167	214	221	243	

This series gives rise to the sequence of plus and minus signs:

$--$	$+$	$-$	$+$	$--$	$+$	$--$	$++$	$--$	$+$	$--$	$++$	$--$	$+++$
1	2	3	4	5	6	7	8	9	10	11	12	13	14

which contains 14 runs.

The test proceeds:

H_0: Sample is random.

H_1: Sample is not random.

(ii) $a = 0.05$.

(iii) Test statistic: $T = \dfrac{R - \frac{1}{3}(2n - 1)}{\sqrt{\{\frac{1}{90}(16n - 29)\}}}$.

This has a standard normal distribution if H_0 is true.

A large number of runs indicates a systematically oscillating series, whereas a small number of runs indicates a steady increase or decrease in sales. A two-tailed test is therefore appropriate, and we reject H_0 if $T \geqslant |1.96|$.

(iv) Sample results:

$$T = \frac{14 - 1/3(2.25 - 1)}{\sqrt{\{1/90(16.25 - 29)\}}} = \frac{-2.33}{2.04} = -1.142.$$

(v) H_0 is not rejected.

Note that in order to detect a trend (i.e., a change in mean value over time) a lower-tailed test would be necessary, since such a trend would be evidenced by a small number of runs. If the salesman is suspected of selling for another organisation on alternate days, resulting in a systematically erratic performance, an upper-tailed test would be needed.

2. Median test of randomness. A second test for randomness uses the number of runs, R, above and below the median of the series as a test statistic. If the series is random it can be shown that R is approximately normally distributed with mean $(n + 2)/2$, and variance $n(n - 2)/\{4(n - 1)\}$. Again the approximation improves as n increases.

Referring once more to our salesman, his median sales are £201. Denoting by A and B sales above and below the median respectively, we have the sequence:

A	BB	A	B	A	BB	AA	BB
1	2	3	4	5	6	7	8

A	M	B	AA	BB	A	BB	AAA
9	10	11	12	13	14	15	

The number of runs, R, is 15.

The test now proceeds along the lines:

(i) H_0: Sample is random.

H_1: Sample is not random.

(ii) $a = 0.05$.

(iii) Test statistic: $T = \dfrac{R - \frac{1}{2}(n + 2)}{n(n - 2)/\{4(n - 1)\}}$.

This has a standard normal distribution if H_0 is true. As before, a large number of runs indicates a systematic oscillation in the series, whilst a small number indicates a tendency for the series to increase or decrease.

A two-tailed test is again appropriate, and H_0 will be rejected if $T > |1.96|$.

(iv) Sample results: $T = \dfrac{15 - \frac{1}{2}(27)}{\sqrt{\{25.23/96\}}} = \dfrac{1.5}{2.447} = 0.613$.

(v) H_0 is not rejected.

Exercises on Section 2

1. The maintenance costs in a machine-shop are expressed as percentages of the total machine-shop costs. Over the last 30 weeks, these percentages have been (in order):

1·84	3·22	1·69	2·55	8·11	9·78	3·25	4·43	7·26	1·14
0·02	8·28	5·58	6·23	3·81	6·15	4·93	3·97	7·99	0·31
0·09	1·14	1·60	9·94	2·20	3·03	1·90	0·14	1·92	3·73

You suspect that maintenance standards are deteriorating, with a consequent increase in costs. Test your hypothesis at the 5% significance level.

2. With the data of Question 1, carry out an alternative test of your suspicions.

3. Tests concerning population median. On occasion, the population median is of interest, since it has the property that it is equally likely to be exceeded or not reached. In addition, the average deviation of the population values from the median is always smaller than from any other figure.

The probability is just $\frac{1}{2}$ that, in sampling from any population, a particular sampled item is above (or below) the median value of the population. In a sample of size n the probability, $P(R = r)$, that the number of such members, R, is just r is therefore given by the binomial distribution:

$$P(R = r) = \frac{n!}{(n-r)!\, r!} (1/2)^{n-r} (1/2)^r$$

$$= \frac{n!}{(n-r)!\, r!} (1/2)^n.$$

Knowledge of the distribution of R enables us to construct tests concerning the population median value.

Consider an insurance company, interested in relating premiums to the median of the claims, since its rates are then, on average, as close as possible to the actual claims.

In particular, the company takes a sample of 15 claims made under the windscreen replacement clause of its standard motor policy:

Value of Claims (£)

25·30	37·80	52·10
21·16	43·18	31·27
49·27	55·76	34·87
39·24	27·19	20·72
41·61	33·47	35·61

Now,

$$P(R = 0) = 0.00003 = P(R = 15)$$
$$P(R = 1) = 0.00046 = P(R = 14)$$
$$P(R = 2) = 0.00320 = P(R = 13)$$
$$P(R = 3) = 0.01389 = P(R = 12)$$
$$P(R = 4) = 0.04166 = P(R = 11)$$
$$P(R = 5) = 0.09164 = P(R = 10)$$
$$P(R = 6) = 0.15274 = P(R = 9)$$
$$P(R = 7) = 0.19638 = P(R = 8)$$

Thus,

$$P(R \leqslant 1) = 0.00049 = P(R \geqslant 14)$$
$$P(R \leqslant 2) = 0.00369 = P(R \geqslant 13)$$
$$P(R \leqslant 3) = 0.01758 = P(R \geqslant 12)$$
$$P(R \leqslant 4) = 0.05924 = P(R \geqslant 11)$$

and therefore,

$$P(R \leqslant 1 \text{ or } R \geqslant 14) = 0.00098$$
$$P(R \leqslant 2 \text{ or } R \geqslant 13) = 0.00738$$
$$P(R \leqslant 3 \text{ or } R \geqslant 12) = 0.03516$$
$$P(R \leqslant 4 \text{ or } R \geqslant 11) = 0.11848$$

Armed with these probabilities we are now in a position to conduct tests concerning the population median claim.

Let us suppose that the company sets its premiums on the assumption of a median value of £44·00. To test this assumption set:

(i) H_0: median = 44·00.
        H_1: median $\neq$ 44·00.

(ii) $a = 0.05$.

(iii) Test statistic: Number, R, of members of sample below (or above) median.

If H_0 is true, median is 44·00. Critical values of R are 3 and 12. We reject H_0 if $R \leqslant 3$ or $R \geqslant 12$, since such values of R occur with a probability of only 0·03516 if H_0 is true.

(iv) Sample results:

$R = 12$, since 12 members of the sample are less than 44·00.

(v) Reject H_0.

Notice that since $P(R \leqslant 2 \text{ or } R \geqslant 13) = 0.00738 < 0.01$, we cannot reject H_0 at the 1 % significance level.

4. Tests of goodness of fit. To complete our brief look at some of the non-parametric statistics arising in work with one sample, we examine a test designed to detect whether empirical (sample) data are consistent with a preconceived theoretical distribution.

The method centres on the calculation of the numbers of sample members which would be expected to fall in various ranges, if the sample were in fact drawn from a population of an assumed type. These expected frequencies are then compared with the observed frequencies and an χ^2-statistic (*cf.* Chapter 9, Section 3) calculated.

Worked Examples

1. A motor dealer suspects that the demands made on him for a particular model of car occur randomly, and are not affected by his local advertising. His records over the last 50 days' trading show the following pattern:

No. of demands	0	1	2	3	4	5	6	7
No. of days	11	12	10	7	5	3	2	0

and include several days when he took advertising space in the local papers. Do the data bear out his suspicions?

Answer

The average daily demand

$$= \frac{(11 \times 0) + (12 \times 1) + \ldots + (2 \times 6) + (0 \times 7)}{(11 + \ldots + 2 + 0)}$$
$$= \frac{100}{50}$$
$$= 2.$$

The assumption that demands occur at random implies that they follow a Poisson distribution with a mean whose value, λ, is estimated by the average observed daily demand, *i.e.* $\lambda \doteq 2$. On this basis the probability, $P(r)$, that on a particular day there will be r demands is given by

$$P(r) = \frac{e^{-\lambda} \lambda^r}{r!} = \frac{e^{-2} 2^r}{r!}.$$

Furthermore, out of a sample of 50 days, the expected number on which exactly r demands are received will be $50P(r)$. The full analysis is shown in the first four columns of Table 43.

Table 43. Analysis of demand for car—Goodness-of-fit test for Poisson distribution

Number of demands (r)	Observed number of days (O)	Probability, P(r)	Expected number of days (E)	$(O - E)^2$	$(O - E)^2/E$
0	11	0·1353	6·765	17·935	2·651
1	12	0·2706	13·53	2·341	0·173
2	10	0·2706	13·53	12·461	0·921
3	7	0·1804	9·02	4·080	0·452
4	5 ⎫	0·0902	4·51 ⎫		
5	3 ⎬ 10	0·03608	1·804 ⎬ 7·155	8·094	1·131
6	2 ⎪	0·01203	0·602 ⎪		
≥7	0 ⎭	0·00479	0·240 ⎭		
Total	50	1·000	50		5·328

The fifth and sixth columns detail the calculation of the χ^2-statistic, $\Sigma(O - E)^2/E$. Notice that the last 4 categories have been unified, in order that none of the categories used has an expected frequency of less than 5 (*see* Chapter 9, Section 6).

Clearly, a very large value of χ^2 will imply considerable deviation of the data from the supposed Poisson model, and a very small value will indicate a suspiciously high agreement between observed and theoretical frequencies. The full test is conducted as follows:

(i) H_0: Sample from a Poisson distribution.
               H_1: Sample is not from a Poisson distribution.

(ii) $a = 0.05$.

(iii) Test statistic: $\chi^2 = \sum(O - E)^2/E$.

If H_0 is true this statistic will have a χ^2 distribution. To ensure that the distribution with the appropriate number of degrees of freedom is used requires some care.

In Table 43 five categories are used in calculating the statistic, *viz.* 0, 1, 2, 3, ≥4. Two constraints have been imposed: (*i*) the expected frequencies total 50, (ii) the expected frequencies average 2. We are left therefore with $5 - 1 - 1 = 3$ degrees of freedom.

Critical value: $\chi_3^2(0\cdot05) = 7\cdot81$.
(iv) Sample results: $\chi^2 = 5\cdot328$.
(v) H_0 cannot be rejected.

2. The daily costs of a liquid chemical used in a manufacturing process are recorded over a 160-day period, and shown in the first two columns of Table 44.

Is it reasonable to suppose that costs follow a normal distribution?

Answer

The variable in this example is effectively continuous.

The sample mean and sample variance, $\bar{c}$ and s^2, are calculated from the grouped data as:

$$\bar{c} = 1468/160 = 9\cdot175.$$
$$s^2 = \frac{15,800 - (1468)^2/160}{159} = 14\cdot66,$$

i.e.

$$s = 3\cdot829.$$

Using these two estimates of the parameters of the assumed normal distribution of costs, the probability that on a particular day the cost is, for example, greater than or equal to £12, and less than £14, is given by:

$$P(12 \leqslant C < 14) = \Phi\left(\frac{(14 - 9\cdot175)}{3\cdot829}\right) - \Phi\left(\frac{(12 - 9\cdot175)}{3\cdot829}\right)$$
$$= \Phi(1\cdot26) - \Phi(0\cdot74)$$
$$= 0\cdot8962 - 0\cdot7704$$
$$= 0\cdot1258.$$

Proceeding in this way throughout the table, gives the figures in the sixth column of Table 44. To conduct the test:

(i) H_0: Costs have a normal distribution.
          H_1: Costs do not have a normal distribution.

(ii) $a = 0\cdot05$.

(iii) Test statistic: $\chi^2 = \Sigma(O - E)^2/E$.

Under H_0, this has a χ^2 distribution. The number of degrees of freedom is $8 - 3 = 5$, since eight categories are used in the final calculation, and three constraints are imposed, *viz*:

(i) Expected frequencies total 160.
(ii) Mean is $9\cdot175$.
(iii) Variance is $14\cdot66$.

Critical value: $\chi_5^2(0\cdot05) = 11\cdot07$.

(iv) Sample results: $\chi^2 = 5\cdot318$.
(v) H_0 is accepted.

Table 44. Analysis of costs of chemical—Goodness-of-fit test for normal distribution

No. of days (O_i)	Cost/day, C (£)	m_i	$O_i m_i$	$O_i m_i^2$	Probability	Expected no. of days (E_i)	$(O_i - E_i)^2$	$\dfrac{(O_i - E_i)^2}{E_i}$
3 ⎫ 16	C < 2	1	3	3	0·0307	4·912 ⎫ 14·16	3·386	0·239
13 ⎭	2 ⩽ C < 4	3	39	117	0·0578	9·248 ⎭	2·663	0·145
20	4 ⩽ C < 6	5	100	500	0·1148	18·368	16·000	0·571
24	6 ⩽ C < 8	7	168	1176	0·1750	28·000	11·614	0·348
30	8 ⩽ C < 10	9	270	2430	0·2088	33·408	58·860	2·007
37	10 ⩽ C < 12	11	407	4477	0·1833	29·328	26·296	1·306
15	12 ⩽ C < 14	13	195	2535	0·1258	20·128	0·370	0·035
10	14 ⩽ C < 16	15	150	2250	0·0663	10·608	4·000	0·667
8	16 ⩽ C	17	136	2312	0·0375	6·000	—	—
160	—	—	1468	15,800	1·00	160	—	5·318

Exercises on Section 4

3. Over a period of 100 days, during a certain minute of the day, the numbers of phone calls coming into the switchboard of a company were as follows:

No. of calls:	0	1	2	3	4	5	6	7	8
Observed no. of days:	5	7	30	40	7	4	5	1	1

Fit a Poisson distribution to the data, and use a χ^2 test to determine the "goodness of fit." Use a 0·01 level of significance.

4. A store's manager records the demand for Fustycrust loaves over a period of 100 shopping days, with the following results:

Number of loaves demanded	Number of days demanded
100	0
105	3
110	4
115	7
120	7
125	9
130	11
135	12
140	14
145	13
150	10
155	6
160	3
165	0
170	1
175	0

Is it reasonable for him to assume that demand follows approximately a normal distribution?

5. A portfolio of five similar shares is studied over a period of 200 days' trading. The number of shares showing an increase in price is noted each day with the following results:

No. of shares increasing in price	No. of days
0	12
1	28
2	70
3	52
4	36
5	2

Estimating from the data the probability that a share price increases, test the goodness of fit of a binomial distribution to these results.

TESTS INVOLVING TWO MATCHED SAMPLES

5. Sign test. The paired-comparison test introduced earlier (*see* Chapter 8, Section 7) requires the differences between pairs to follow a normal distribution. A non-parametric test, however, can be based on the fact that differences between dependent samples from two populations with equal means will be positive (or negative) with a probability of 1/2. Taking the number of positive differences as the test statistic, a test can be devised using the binomial distribution, rather as in Section 3 above.

Worked Example

A firm of accountants is considering the introduction of one of two self-teaching programs into its training scheme. Both are concerned with sampling in auditing. A number of the firm's senior staff are asked to evaluate the programs on a scale from 1–10. The results are shown in Table 45. Is there any reason to suppose that one program is favoured more than the other?

Table 45. Staff scores for "Samplan" and "Testy"

Staff member	"Samplan"	"Testy"	Difference	
1	6	8	−2	(−)
2	7	9	−2	(−)
3	5	8	−3	(−)
4	6	7	−1	(−)
5	3	2	1	(+)
6	4	5	−1	(−)
7	4	9	−5	(−)
8	1	3	−2	(−)
9	6	3	3	(+)
10	4	6	−2	(−)
11	5	2	3	(+)
12	2	7	−5	(−)
13	9	9	Tie	
14	0	8	−8	(−)
15	5	1	4	(+)
16	4	5	−1	(−)
17	1	2	−1	(−)
18	2	4	−2	(−)
19	7	7	Tie	
20	3	6	−3	(−)
21	8	3	5	(+)
22	4	9	−5	(−)
23	6	8	−2	(−)
24	2	2	Tie	
25	9	1	8	(+)

Answer

There are three ties in the table. We could assume the scores to be on a continuous scale (in which case the probability of two being alike is zero), and decide on the sign by the toss of a coin. Alternatively, since the sample is fairly large, and the number of ties small, they could be simply ignored in the analysis. We shall take the latter course.

(i) H_0: Testy and Samplan equally good.
            H_1: Testy better than Samplan.
(ii) $a = 0 \cdot 05$.
(iii) Test statistic: $T =$ number of minus differences.

If H_0 is true, using the binomial distribution with $n = 22$ (*i.e.*, $25 - 3$ ties), and $p = q = \frac{1}{2}$, we find that:

$$P(T \geqslant 15) = 0 \cdot 0669.$$
$$P(T \geqslant 16) = 0 \cdot 0262.$$
$$P(T \geqslant 17) = 0 \cdot 0085.$$

Thus H_0 is rejected in favour of H_1 if $T \geqslant 16$.

(iv) Sample results: $T = 16$.
(v) Reject H_0 at 5% significance.

(Note that $T \geqslant 17$ for H_0 to be rejected at $a = 1\%$.)

It is interesting to note that this test does not require any assumptions to be made about the populations, which need not have the same distributions. Additionally the test may be used for scores.

6. Wilcoxon Signed Rank-Sum test. The disadvantage of the previous test is that, although it takes account of the signs of differences, it makes no allowance for their magnitudes.

Under the assumption that there is no difference between the population means, it can be shown that for a "large" sample ($n \geqslant 10$) the statistic $T =$ "sum of ranks of less frequent sign" has approximately a normal distribution with mean $n(n + 1)/4$, and variance $n(n + 1)(2n + 1)/24$. Special tables have been prepared giving the distribution of T for small samples (*see* for example, Bibliography, p. 298).

Worked Example

Two models of machine are under consideration for purchase. A company has one of each type for a trial period, and a team of 14 operators use each machine for a fixed length of time. Is there any reason to suppose, using the data of Table 46, that there is any significant difference between the output capacities of the two machines?

Table 46. Wilcoxon Signed Rank-Sum Test using machine output data

Operator	Output from machine A	Output from machine B	Difference in output	Rank of difference	Ranks with less frequent sign
1	78	60	−18	11	
2	75	58	−17	10	
3	53	46	−7	7	
4	68	71	+3	3	→ 3
5	82	80	−2	2	
6	64	59	−5	5	
7	95	73	−22	13	
8	86	78	−8	8	
9	64	37	−27	14	
10	71	75	+4	4	→ 4
11	54	60	+6	6	→ 6
12	80	79	−1	1	
13	51	38	−13	9	
14	70	51	−19	12	

Total $(T) = 13$

Answer

The fourth column of Table 46 gives the (signed) difference between an operator's performance on the two machines. These differences are ranked (irrespective of sign) in the fifth column. There are fewer differences of positive sign (3) than of negative (11). Our statistic, T, is found therefore by summing the ranks of the differences with positive signs.

(i) H_0: Average machine outputs equal.
        H_1: Average machine outputs unequal.

(ii) $a = 0.05$.

(iii) Test statistic: $\dfrac{T - \frac{n}{4}(n+1)}{\sqrt{\{n(n+1)(2n+1)/24\}}}$.

This has a standard normal distribution if the null hypothesis is true. Critical values of T: ± 1.96.

(iv) Sample results:

$$\frac{T - \frac{n}{4}(n+1)}{\sqrt{\dfrac{n(n+1)(2n+1)}{24}}} = \frac{13 - \dfrac{14.15}{4}}{\sqrt{\dfrac{14.15.29}{24}}} = \frac{-39.5}{15.93} = -2.48.$$

(v) H_0 is rejected at $a = 0.05$ (but not at $a = 0.01$, when critical value is ± 2.58).

Exercises on Section 6

6. In Question 10 of Exercises on Section 7, Chapter 8 use a non-parametric statistic to test whether achievements against budgets are slipping.

7. Perform a non-parametric test using the data of Question 11, Exercises on Section 7, Chapter 8.

8. Use a non-parametric statistic to test the costliness of Castor *v.* Pollux in Question 12, Exercises on Section 7, Chapter 8.

7. Rank correlation coefficient*. If the degree of correlation is required between two variables which do not lend themselves to precise measurement (such as managerial ability, creativity, beauty, etc.) the coefficient defined in Chapter 11, Section 7 is of no use. Nor indeed do conclusions concerning correlation remain valid if assumptions of normality are dropped.

The *rank correlation coefficient*, *r*, overcomes both these limitations, and is defined by:

$$r = 1 - \frac{6 \, \Sigma d^2}{n(n^2 - 1)}$$

where *n* is the number of members of each sample, and *d* is the difference in ranks between a matched pair of sample members. Furthermore, tests concerning the degree of correlation can be based on the fact that for large *n* ($\geqslant 10$), the statistic

$$\frac{r\sqrt{(n - 2)}}{\sqrt{(1 - r^2)}}$$

has approximately a *t*-distribution with ($n - 2$) degrees of freedom.

Worked Example

Two managers are asked to rank a group of employees in order of potential for eventually becoming top managers. The rankings are shown in the second and third columns of Table 47.

Calculate the rank correlation coefficient between the two managers' assessments.

Answer

The differences between ranks, *d*, and d^2 are given in the last two columns of Table 47. The sample sizes are $n = 20$.

$$\text{Rank correlation coefficient} = 1 - \frac{6 \times 214}{20(20^2 - 1)}$$
$$= 0{\cdot}8391.$$

* This section should preferably be omitted until after reading Chapter 11.

Table 47. Rank correlation between managers

Employee	Ranking by manager I	Ranking by manager II	Difference between rank (d)	d²
A	1	3	2	4
B	2	4	2	4
C	3	6	3	9
D	4	2	−2	4
E	5	1	−4	16
F	6	5	−1	1
G	7	9	2	4
H	8	8	0	0
I	9	7	−2	4
J	10	13	3	9
K	11	10	−1	1
L	12	19	7	49
M	13	17	4	16
N	14	16	2	4
O	15	15	0	0
P	16	14	−2	4
Q	17	11	−6	36
R	18	18	0	0
S	19	12	−7	49
T	20	20	0	0
			Total	214

Exercises on Section 7

9. Calculate the rank correlation coefficient between the budgeted variances for 1976 and 1977 in Question 10, Exercises on Section 7, Chapter 8.

10. The following data gives costs and sales for 12 supermarkets. Calculate the rank correlation coefficient and test for the significance of the correlation.

Supermarket	Costs	Sales
1	22	38
2	20	30
3	28	39
4	26	30
5	24	31
6	41	68
7	43	64
8	29	35
9	45	59
10	35	43
11	37	47
12	33	42

TESTS INVOLVING TWO INDEPENDENT SAMPLES

8. Median test. Suppose we are interested in whether two independent samples have been drawn from the same population or not. If indeed they have, we would expect that roughly half the members of each sample would be above, and half below, the median value of the combined sample. A test is easily developed based on the observed numbers above and below the overall median.

Worked Example

An insurance company is interested in knowing whether there is any significant difference in claims made by male and by female policy-holders. Samples of 14 "male" and 16 "female" claims are taken:

Male (£)	Female (£)
62	93
38	101
43	72
79	118
77	100
23	45
11	68
52	72
33	47
41	83
70	92
49	106
69	63
43	66
	85
	81

Is there any reason to suppose the populations different?

Answer

The median of the combined sample of 30 claims is 68·5. The observed numbers in the two samples above and below the median are tabulated below, with the expected numbers in brackets.

	Male	Female	Total
Above median	4 (7)	11 (8)	15
Below median	10 (7)	5 (8)	15
Total	14	16	30

The χ^2-statistic for this contingency table can now be calculated (remembering the continuity correction since we have only

$$(2 - 1) \times (2 - 1) = 1 \text{ degree of freedom}):$$

$$\chi^2 = \frac{\{|4 - 7| - \frac{1}{2}\}^2}{7} + \frac{\{|11 - 8| - \frac{1}{2}\}^2}{8} + \frac{\{|10 - 7| - \frac{1}{2}\}^2}{7}$$

$$+ \frac{\{|5 - 8| - \frac{1}{2}\}^2}{8}$$

$$= 3 \cdot 35.$$

The full test is therefore:

    (i) H_0: Samples are drawn from identical population.
             H_1: Samples are not drawn from identical population.
    (ii) $a = 0 \cdot 10$.
   (iii) Test statistic: χ^2.
       This is distributed as χ_1^2 if H_0 is true.
       Critical value of χ^2 is $2 \cdot 71$.
   (iv) Sample results: $\chi^2 = 3 \cdot 35$.
    (v) Reject H_0.

9. Wald-Wolfowitz number of runs test.

An alternative procedure is to arrange the combined figures sampled from the two populations in ascending order of magnitude, and to count the number of uninterrupted runs of members of the same sample.

Under the null hypothesis that the two populations are the same, a large number of runs R (indicating an "interleaving" of samples) would be expected. In fact, if the combined sample is greater than about 20, it can be shown that (approximately),

$$\frac{R - \left(\dfrac{2n_1 n_2}{n_1 + n_2} + 1\right)}{\sqrt{\left\{\dfrac{2n_1 n_2}{(n_1 + n_2)^2} \cdot \dfrac{(2n_1 n_2 - n_1 - n_2)}{(n_1 + n_2 + 1)}\right\}}} \sim N(0, 1)$$

where n_1, n_2 are the sizes of the two samples.

For a combined sample size of less than 20 it is necessary to consult special tables (*see* Bibliography, p. 298).

Worked Example

Using the data given in the worked example of Section 8 above apply the Wald-Wolfowitz test.

Answer

Arranging the 30 claims figures in ascending order; and denoting their sample of origin by M (male) and F (female):

11	23	33	38	41	43	43	45	47	49
M	M	M	M	M	M	M	Γ	Γ	M

52	62	63	66	68	69	70	72	72	77
M	M	F	F	F	M	M	F	F	M

79	81	83	85	92	93	100	101	106	118
M	F	F	F	F	F	F	F	F	F

The number of runs, R, in this sequence is 8. Thus:

(i) H_0: Samples come from same population.
        H_1: Samples do not come from same population.

(ii) $a = 0.05$.

(iii) Test statistic: R.

If H_0 is true R has a normal distribution with mean

$$= \frac{2.14.16}{14 + 16} + 1 = 15.93$$

and variance

$$= \frac{2.14.16}{(14 + 16)^2} \cdot \frac{(2.14.16. - 14 - 16)}{(14 + 16 + 1)}$$
$$= 6.712.$$

A lower-tailed test is appropriate since a small number of runs discredits H_0. The critical value for R is therefore

$$15.93 - 1.64 \sqrt{(6.712)} = 11.68.$$

(iv) Sample results: $R = 8$.
(v) H_0 is rejected.

Exercises on Section 9

11. Contracts for fuel to run a manufacturer's fleet of delivery vehicles are to be placed with one of two petrol companies. Tests on the fuel each company recommends reveal the following costs/mile, for two samples of the vehicles:

Fuel I	5·1	2·7	4·2	2·7	4·1	2·9	2·7	2·3	2·5	3·5
	2·1	3·7	3·7							
Fuel II	3·6	1·9	3·1	1·9	2·5	1·6	3·9	2·5	2·1	1·7

Perform a sign test for difference between the populations.
12. Using the data of Question 11, carry out the Wald-Wolfowitz test.

CHAPTER 11

REGRESSION AND CORRELATION ANALYSIS

INTRODUCTION

We have already (p. 50) discussed situations where one variable may be dependent on another. We will now take our examination of dependence much further. In *regression analysis* we are concerned with deriving an exact mathematical relationship which attempts to explain the dependence of one set of numerical quantities on another, and in *correlation analysis* we provide a measure of the degree of dependence between the two sets. Both regression and correlation analyses form part of the study of *association* between variables.

We are frequently presented with two (or indeed more than two) associated variables. For example, we may have details of man-hours worked and output, of advertising expenditure and sales revenue, or of total costs and level of production. One of the main objectives in examining the association between variables is to enable the prediction of one value when the other is given. Thus we could estimate future output if we knew the number of man-hours which would be available, or we could estimate the effect on sales revenue of investing a certain amount of money in advertising.

In the following sections we shall concern ourselves first with regression analysis.

SIMPLE LINEAR REGRESSION

1. General comments. There are innumerable mathematical relationships which could exist between two variables. Of these the simplest category is that of the *linear* function, which is characterised by the equation

$$Y = mX + c$$

where Y and X are variables, and m and c are constants. Thus,

$$Y = 4X + 6,$$
$$Y = 3 \cdot 5X + 7 \cdot 2,$$
$$Y = 13 \cdot 8X + 76 \cdot 1,$$

are all examples of linear functions. Each linear function has a graphical straight-line representation, hence the name.

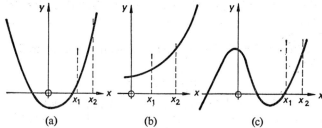

FIG. 71.—Some possible non-linear relationships between x and y. In each case the relationship is approximately linear over the range (x_1, x_2).

Because of its relative simplicity, a straight-line relationship is often assumed between two variables. This may seem an oversimplification, but there is little point in assuming a more complex relationship at the outset, unless we have strong grounds for doing so. Indeed, for any relationship, if we take a small range of values of X the dependence of Y on X follows approximately a straight line (*see* Fig. 71).

2. Least squares regression line. Let us assume that we are presented with sets of corresponding values of two variables (X, Y). Plotting the pairs of values as points on a graph results in a *scatter diagram* (Fig. 72).

Presumably our empirical values of X and Y are subject to a certain amount of random error. For example, in the case of man-hours worked and output, working conditions on a particularly hot day, the breakdown of a machine, or the introduction of a relatively inexperienced operative may each result in an output lower than expected. We have reason to suppose from the scatter diagram that the theoretical relationship between X and Y may well be linear, and that deviations from linearity are due to random effects of this type. The immediate problem is to choose, out of the multitude of lines

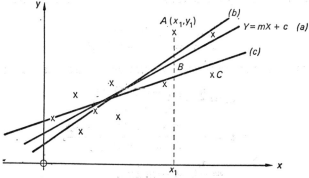

FIG. 72.—Scatter diagram of observed pairs (x, y).

available, the one which fits the data most satisfactorily. Lines (a), (b), and (c) in Fig. 72 all seem at first glance to be suitable candidates for the privilege.

Our criterion for choice is illustrated with reference to Fig. 72. Consider the line

$$Y = mX + c$$

and suppose that our data points are $(x_1, y_1), \ldots, (x_n, y_n)$.

The deviation of point A from the line can be measured by the distance AB, and since A is (x_1, y_1) and B is $(x_1, mx_1 + c)$,

$$AB = y_1 - (mx_1 + c).$$

Proceeding in a similar way, we can arrive at measures of the deviations of each of the remaining $(n - 1)$ points from the line. Some of these deviations will be positive, and some (e.g., for data point C) will be negative.

Since adding these deviations together would result in a considerable amount of cancellation between positive and negative values, we choose as our single measure of the scatter, D, of points about the line,

$$D = \{y_1 - (mx_1 + c)\}^2 + \ldots + \{y_n - (mx_n + c)\}^2.$$

The values of m and c are at our disposal, whilst

$$(x_1, y_1), (x_2, y_2), \ldots, (x_n, y_n)$$

are observed values of X and Y. If we can now choose the values of m, c to be such that D is a minimum, we will have defined the *line of best fit* to the data, in the sense that the "scatter" of points about this line is less than that about any other.

The detailed procedure so to choose m and c involves the use of partial differentiation, with which we need not concern ourselves (*see* Bibliography, p. 298 for complete argument). For D to be minimised m and c must satisfy the following pair of simultaneous equations:

$$\Sigma y_i = nc + m \Sigma x_i$$
$$\Sigma x_i y_i = c \Sigma x_i + m \Sigma x_i^2.$$

These so-called *normal equations* may be solved for m and c to give:

$$m = \frac{n \Sigma x_i y_i - (\Sigma x_i)(\Sigma y_i)}{n \Sigma x_i^2 - (\Sigma x_i)^2}. \tag{i}$$

$$c = \frac{(\Sigma y_i)(\Sigma x_i^2) - (\Sigma x_i)(\Sigma x_i y_i)}{n (\Sigma x_i^2) - (\Sigma x_i)^2}.$$

$$= \frac{1}{n} (\Sigma y_i - m \Sigma x_i). \tag{ii}$$

The line defined by the calculated values of m and c is referred to as the *least squares regression line of Y on X*.

The following example illustrates the fitting in practice of a regression line to a set of data.

Worked Example

Total costs (comprised of fixed and variable costs) are given over an 8-week period for Magwitch Ltd., together with figures for the company's corresponding output of jaggers (*see* Table 48, columns 1–3).

Table 48. Output of jaggers and total costs for Magwitch Ltd.

(1) Week	(2) Output of jaggers (X)	(3) Total costs (Y) in £00	(4) XY	(5) X²	(6) Y²
1	2	11·2	22·4	4	125·44
2	3	15·6	46·8	9	243·36
3	5	20·3	101·5	25	412·09
4	4	20·8	83·2	16	432·64
5	1	7·8	7·8	1	60·84
6	3	10·6	31·8	9	112·36
7	2	12·3	24·6	4	151·29
8	4	21·5	86·0	16	462·25
9	5	22·0	110·0	25	484·00
10	6	27·6	165·6	36	761·76
Total	35	169·7	679·7	145	3246·03

Fit a least squares regression line to the data, and interpret the values of m and c.

Answer

The total costs (Y) are dependent on output level (X).

To facilitate the calculations, Table 48 is augmented by columns 4–6. Referring to the normal equations,

$$m = \frac{(10 \times 679\cdot7) - (35 \times 169\cdot7)}{(10 \times 145) - (35)^2} = 3\cdot811.$$

$$c = \frac{1}{10}\{169\cdot7 - (3\cdot811 \times 35)\} = 3\cdot63.$$

The required regression line is therefore:

$$Y = 3\cdot811X + 3\cdot63.$$

Clearly the values of m (the gradient) and of c (the intercept on the y-axis) represent variable cost/unit and fixed costs respectively.

We are now in a position to estimate total costs for any pre-chosen level of output. Incidentally in using regression as a tool for prediction, the estimates produced by *interpolation* are likely to be fairly reliable. However, *extrapolation* can lead to grave difficulties. We are not to know, for example, that beyond the level of output of 6, total costs do not rise as indicated by the dashed line of Fig. 73. Indeed, this general shape of cost curve is quite common.

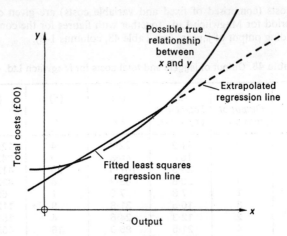

FIG. 73.—Illustration of dangers of extrapolation.

3. The sampling nature of regression analysis. It would be wrong to think that the values of m and c calculated from the normal equations specify the theoretical linear relationship assumed between X and Y. We have already hinted that the values of the dependent variable (Y) are subject to random error. In other words, for a given value of the independent variable (X), the observed Y-value is only one from a population of possible Y-values having an associated probability distribution. For a given value of X, Y is really a random variable from which we take a sample of size one. In our example, if we were to take another ten-week period we would typically obtain a different set of Y-values, corresponding to the same set of X-values.

In the absence of other information we assume that the distribution of Y for a given value of X is normal, with an expected value of $MX + C$. Symbolically,

$$\text{E}(Y|X) = MX + C.$$

For a particular value of X, we will sometimes obtain a value of Y greater than $MX + C$, and sometimes less. On average, however, the value of Y will be $MX + C$.

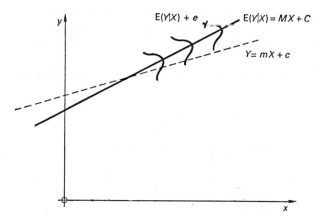

FIG. 74.—True and estimated least squares regression lines.

The situation is made clear by reference to Fig. 74. Note that

(*i*) the "true" regression line is

$$E(Y|X) = MX + C.$$

(*ii*) The observed values of Y can be represented in the form

$$Y = E(Y|X) + e$$

where e is the random error, expressing the deviation of Y (for a given X) from its expected value. It is assumed that the distributions of Y for given values of X (and therefore the random error terms) are all independent of each other, and that their variances are all equal. In fact var $e = \sigma^2$ is known as the *variance of the regression*.

(*iii*) Using observed values of Y for given values of X, the "true" regression line is estimated by

$$Y = mX + c.$$

We now see that the calculated values of m and c are only estimates of M and C, and as such are subject to sampling error.

4. Standard error of the regression. The variance σ^2 (or standard deviation σ) of the individual dependent random variables, Y, cannot be determined exactly. However, after the regression line has been fitted an estimate s^2 can be found from the sum of the squared deviations from the line, *i.e.*

$$s^2 = \sum \frac{\{y_i - (mx_i + c)\}^2}{(n-2)}.$$

This estimate measures the "average" squared deviation from the line, although a divisor of $(n - 2)$ is necessary because two of the n degrees of freedom are sacrificed by estimating M and C from the n data values.

In fact from a computational point of view the alternative (and exactly equivalent) expression

$$s^2 = \frac{\Sigma y_i^2 - m \Sigma y_i - c \Sigma x_i y_i}{n - 2}$$

is more convenient. Referring to the example of Section 2 above.

$$s^2 = \frac{(3246 \cdot 03) - (3 \cdot 81 \times 169 \cdot 7) - (3 \cdot 63 \times 679 \cdot 7)}{10 - 2}$$

$$= \frac{132 \cdot 14}{8}$$

$$= 16 \cdot 52$$

and the *standard error of the regression*, s, is $\sqrt{(16 \cdot 52)} = 4 \cdot 0645$.

It is interesting to note that when we minimise the scatter, D, to obtain the normal equations for m and c (Section 2 above), we are effectively choosing a line which provides us with the minimum value for our estimate s^2 of the variance of the regression.

5. Sampling distributions of estimators of regression parameters. Since m and c only estimate the true regression parameters M, C they both have sampling distributions. It can be shown that the estimators are unbiased, *i.e.*

$$E(m) = M.$$
$$E(c) = C.$$

Furthermore the sampling variances s_m^2 and s_c^2 of m and c are closely related to the standard error of the regression, s^2:

$$s_m^2 = \frac{s^2(\Sigma x_i^2)/n}{\Sigma x_i^2 - (\Sigma x_i)^2/n}$$

and

$$s_c^2 = \frac{s^2}{\Sigma x_i^2 - (\Sigma x_i)^2/n}$$

(for the derivation of the above *see* references in the Bibliography, p. 298).

Finally, the assumption of approximate normality in the distributions of random errors leads to:

$$\frac{m - M}{s_m} \sim t_{n-2} \text{ and } \frac{c - C}{s_c} \sim t_{n-2}$$

for small samples, and

$$\frac{m - M}{s_m} \sim N(0, 1) \text{ and } \frac{c - C}{s_c} \sim N(0, 1)$$

for large samples.

Knowledge of these distributions enables us to construct confidence intervals for C and M, and to test hypotheses concerning the values of C and M in the usual way.

Worked Example

1. Using the data of Section 2 above, find a 90% confidence interval for variable costs.

Answer

The standard error of the regression has been derived already using this data in Section 4 above.

$$s_m{}^2 = \frac{(16 \cdot 52)\dfrac{(145)}{10}}{145 - (35)^2/10} = 10 \cdot 6462$$

i.e.,

$$s_m = 3 \cdot 2629.$$

Now,

$$\frac{m - M}{s_m} \sim t_{n-2} = t_8$$

and 90% of the distributive t_8 lies between $\pm 1 \cdot 86$.
Thus,

$$3 \cdot 81 - (3 \cdot 26 \times 1 \cdot 86) < M < 3 \cdot 81 + (3 \cdot 26 \times 1 \cdot 86)$$

i.e.,

$$2 \cdot 26 < M < 9 \cdot 88.$$

Clearly we cannot be confident about our estimate of M since it is based on so few observations.

2. Using the data of Section 2 above, test the hypothesis that fixed costs are £6. Use a 5% significance level.

Answer

(i)
$$H_0: C = 6.$$
$$H_1: C \neq 6.$$

(ii)
$$\alpha = 0 \cdot 05.$$

(iii) Test statistic:

$$T = \frac{C - 6}{s_c} \sim t_8 \text{ if } H_0 \text{ is true.}$$

Critical value of T is $2 \cdot 31$.
Reject H_0 if $T > 2 \cdot 31$ or $T < -2 \cdot 31$.

(*iv*) Sample results:

$$T = \frac{3\cdot63 - 6}{\sqrt{(16\cdot52/22\cdot5)}} = -2\cdot766.$$

(*v*) Reject H_0.

6. Confidence in predicted values.
Viewing regression analysis from the viewpoint of prediction, two quantities are of immediate interest. In the first instance it is useful to estimate the expected (or average) value of Y, given a particular value of X, *i.e.* $E(Y|X)$. Secondly, we may require an estimate simply of Y, given a particular value of X.

The regression line will provide identical estimates of these two values. Referring, for example, to Section 2 above, when $x = 3$, our estimates are both $(3 \times 3\cdot81) + 3\cdot63 = 15\cdot06$.

The sampling errors involved in making these two estimates are, however, quite different. Denoting by $\bar{Y}_x$ the estimate of $E(Y|X)$, and by $s_{\bar{Y}_x}^2$ its sampling variance, it can be shown that

$$s_{\bar{Y}_x}^2 = s\sqrt{\left\{\frac{1}{n} + \frac{(x - \bar{x})^2}{(\Sigma x^2 - n\bar{x}^2)}\right\}}.$$

Similarly, if $s_{Y_x}^2$ represents the sampling variance of Y, given X then,

$$s_{Y_x} = s\sqrt{\left\{\frac{1}{n} + \frac{(x - \bar{x})^2}{(\Sigma x^2 - n\bar{x}^2)} + 1\right\}}.$$

When n is large these two expressions simplify considerably to the approximate forms:

$$s_{\bar{Y}_x} = s/\sqrt{n}$$

and

$$s_{Y_x} = s.$$

Finally, the statistics

$$\frac{\bar{Y}_x - E(Y|X)}{s_{\bar{Y}_x}} \text{ and } \frac{Y - \bar{Y}_x}{s_{Y_x}}$$

can be shown to both have *t*-distributions with $(n - 2)$ degrees of freedom when n is small, and to have approximately standard normal distributions when n is large.

These admittedly rather unwieldy results are applied quite simply.

Worked Example

Using again the data of Section 2 above, find 90% confidence intervals for (*i*) Y, given $X = 3$, and (*ii*) expected value of Y, given $X = 3$.

Answer

(i)
$$S_{Y_x} = \sqrt{\left\{\frac{1}{10} + \frac{1}{90} + 1\right\}} \sqrt{(16 \cdot 52)}$$

$$= 4 \cdot 2837.$$

Thus,

$$15 \cdot 06 - (1 \cdot 86 \times 4 \cdot 2837) < Y < 15 \cdot 06 + (1 \cdot 86 \times 4 \cdot 2837).$$

i.e.,

$$7 \cdot 092 < Y < 23 \cdot 028.$$

(ii)
$$S_{\bar{Y}_x} = \sqrt{\left\{\frac{1}{10} + \frac{(3 - 3 \cdot 5)^2}{22 \cdot 5}\right\}} \sqrt{(16 \cdot 52)} = 1 \cdot 3548.$$

90% of the area of the t-distribution with $10 - 2 = 8$ degrees of freedom lies between $\pm 1 \cdot 86$.

Thus,

$$15 \cdot 06 - (1 \cdot 86 \times 1 \cdot 3548) < E(Y|X = 3) < 15 \cdot 06 + (1 \cdot 86 \times 1 \cdot 3548)$$

i.e.,

$$12 \cdot 54 < E(Y|X = 3) < 17 \cdot 58.$$

Exercises on Section 6

1. A department store chain collected data on sales wages and sales volume for each of its stores. These data for the past month are:

Store	Sales wages (£000)	Sales volume (£00,000)
A	50·3	4·9
B	31·6	3·2
C	40·2	4·7
D	63·7	5·6
E	22·6	2·7
F	46·4	4·8
G	40·8	3·7
H	39·7	3·6
	335·3	33·2

(a) Compute the least squares line of regression for the data.

(b) Compute a point estimate of sales wages, if sales volume is £420,000.

(c) What is the 95% confidence interval estimate of sales wages if sales volume is £640,000?

(d) Give a 95% confidence interval for expected sales wages when sales volume is £350,000.

2. A record of maintenance cost is kept on each of 14 nearly identical automatic machines. These data are to be compared with machine age to

determine if a suitable functional relationship can be derived for estimating maintenance cost.

Maintenance cost (£)	Age (years)
124	6
47	2
190	7
60	5
115	3
20	1
95	6
63	2
80	4
60	1
108	8
112	5
151	9
39	3

(a) Fit a least squares regression line to the data.

(b) Compute a point estimate for the maintenance cost for a machine five years of age.

(c) Find the 95% interval estimate of maintenance cost for a machine three years of age.

3. Data is collected for the cost (x_1) of packaging and the package volume (x_2):

Cost (x_1) (pence)	Volume (cubic feet)
10·8	0·72
14·4	0·88
19·6	1·23
18·0	1·12
8·4	0·48
15·2	1·09
11·0	0·63
13·3	0·82
23·1	1·40

(a) Estimate the regression of cost on package volume.

(b) Compute the estimated standard error of the regression.

(c) Compute the 95% interval estimate of the mean value of cost for a package volume of 1·15 cubic feet.

(d) What is the 95% interval estimate of cost for a package volume of 0·80 cubic feet?

4. The following are total investments of the Ruritanian Banks from 1965 to 1974, where t represents the years starting with $1965 - 1$.

Year	Investments ($ billion)	Year
1965	4·3	1
1966	4·0	2
1967	3·9	3
1968	4·0	4
1969	4·0	5
1970	4·5	6
1971	3·8	7
1972	3·4	8
1973	3·4	9
1974	4·1	10

(a) Find the least squares trend line.
(b) Find the standard error of the regression.

5. The manager of a motor company assumes that demand for his products depends on advertising expenditures. The following 1974 sales data are available for ten regions:

Region	Sales (000/year)	Advertising (£000)
A	1	1
B	1	0
C	1	2
D	2	0
E	2	1
F	3	3
G	4	4
H	4	2
I	5	4
J	5	3

Find

(a) the regression line of sales on advertising;
(b) the estimated values of sales;
(c) the standard error of these estimates.

6. Suppose now the manager assumes that demand for his product depends on the disposable income of each region. As a forecasting device, he decides to use simple linear regression analysis. Given the following data:

Region	Sales (Y) (000/year)	Disposable income (X) (£000,000)
A	1	10
B	1	20
C	1	10
D	2	30
E	2	20
F	3	40
G	4	30
H	4	50
I	5	60
J	5	60

Find

    (a) regression line of Y on X;

    (b) estimated values of sales;

    (c) a 95% confidence interval for the expected sales when disposable income is £35,000,000.

CORRELATION

7. Sample coefficient of determination. We now turn to the second half of our discussion of association and concern ourselves with measuring the degree of dependence between variables. This connects very simply with our previous work on covariance.

In essence, the measurement of dependence with regard to two variables is based on a determination of the proportion of the total dispersion amongst the Y-values which can be explained by the dependence of Y on X. Consider the scatter diagrams of Fig. 75.

With the usual notation the total dispersion amongst the Y-values is measured in each case by $\Sigma(y - \bar{y})^2$. In Fig. 75 (a) all of this dispersion is attributable to the dependence of Y on X. In other words, given a value of X there is no doubt about the resulting value for Y. The situation in Fig. 75 (b) is, however, quite different. There, although there is a general tendency for Y to increase as X increases,

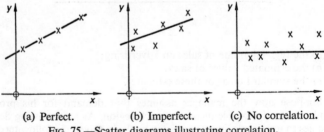

     (a) Perfect.   (b) Imperfect.   (c) No correlation.

FIG. 75.—Scatter diagrams illustrating correlation.

not all of the dispersion amongst the Y-values is due to changes in X alone. Finally, in Fig. 75 (c) there is no tendency for increasing (or decreasing) values of X to be associated with increasing (or decreasing) values of Y.

The key concept is a realisation that the scatter of points about the regression line is a measure of the degree to which the variation in Y-values has not been explained by changing X-values.

The deviation of any Y-value, y, from the mean, $\bar{y}$, can be analysed into two components:

$$y - \bar{y} = (y - y_R) + (y_R - \bar{y})$$

where (x, y_R) is a point on the regression line (see Fig. 76).

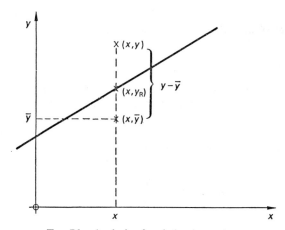

FIG. 76.—Analysis of variation in y-values.

The two components represent respectively the parts of the deviation unexplained and explained by the dependence of Y on X. Now, it can be shown* that:

$$\Sigma(y - \bar{y})^2 = \Sigma(y - \bar{y}_R)^2 + \Sigma(y_R - \bar{y})^2.$$

The three terms in this equation can be interpreted respectively as the total dispersion amongst the Y-values, the dispersion not due to dependence on X, and the dispersion due to dependence on X. A

$$
\begin{aligned}
{}^*\Sigma(y - \bar{y})^2 &= \Sigma\{(y - y_R) + (y_R - \bar{y})\}^2 \\
&= \Sigma(y - y_R)^2 + 2\Sigma(y - y_R)(y_R - \bar{y}) + \Sigma(y_R - \bar{y})^2.
\end{aligned}
$$

Now

$$
\begin{aligned}
\Sigma(y - y_R)(y_R - \bar{y}) &= \Sigma\{y - (mx + c)\}\{(mx + c) - \bar{y}\} \\
&= \Sigma\{y - (mx + c)\}\{c - \bar{y}\} + \Sigma\{y - (mx + c)\}mx \\
&= \Sigma\{xy - (mx^2 + cx)\}m \\
&= 0.
\end{aligned}
$$

natural measure of the association between X and Y is therefore provided by the *sample coefficient of determination, r^2* defined by:

$$r^2 = \frac{\text{dispersion in } Y\text{-values due to dependence on } X}{\text{total dispersion in } Y\text{-values}}$$

$$= \Sigma(y_R - \bar{y})^2/\Sigma(y - \bar{y})^2$$

$$= 1 - \frac{\Sigma(y - y_R)^2}{\Sigma(y - \bar{y})^2}.$$

Referring again to Fig. 75,

(*a*): $y - y_R = 0$ for all points,

$$\therefore \quad r^2 = 1.$$

(*c*): $y_R = \bar{y}$ since there is no tendency for Y to increase or decrease with X

$$\therefore \quad r^2 = 0.$$

(*b*): This is the typical intermediate situation where r^2 lies between 0 and 1. It should be clear from its definition as a proportion, that r^2 is always positive and never exceeds one. A value of r^2 near one indicates a high degree of association between the variables, whereas a value near zero implies little association.

The square root of the sample coefficient of determination, r, is itself used as a measure of association and is known as the *sample coefficient of correlation*.

Worked Example

Find the sample coefficient of determination between total costs and production level using the data from the worked example of Section 2 above.

Answer

$$r^2 = 1 - \frac{\Sigma(y - y_R)^2}{\Sigma(y - \bar{y})^2}$$

$$= 1 - \frac{\Sigma(y - y_R)^2}{\Sigma y^2 - n\bar{y}^2}.$$

The quantity $\Sigma(y - y_R)^2 = 40\cdot36$ has already been found in the process of calculating the standard error of the regression in Section 4 above. Thus,

$$r^2 = 1 - \frac{132\cdot14}{3246\cdot03 - 10(16\cdot97)^2}$$

$$= 1 - \frac{132\cdot14}{366\cdot22}$$

$$= 0\cdot6392.$$

This high degree of association is supported by the small scatter of points about the regression line.

8. Alternative approach to measurement of association. The definition of r^2 that has been given does not lend itself to easy evaluation. Let us therefore approach the measure of association from a different angle.

Consider the statistic, C, defined by:

$$C = \Sigma(x - \bar{x})(y - \bar{y}).$$

If values of Y above (below) the mean, $\bar{y}$, tend to arise from values of X above (below) the mean $\bar{x}$ then the products in the summation will be predominantly positive, and therefore C will be positive. Conversely, a general correspondence of Y-values above (below) $\bar{y}$, with X-values below (above) $\bar{x}$ will lead to a negative value for C. No tendency for correspondence either way will give rise to some negative and some positive terms in the sum, which will therefore be small. The three possibilities are illustrated in Fig. 77.

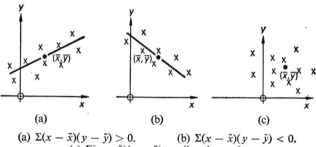

(a)   (b)   (c)

(a) $\Sigma(x - \bar{x})(y - \bar{y}) > 0.$ (b) $\Sigma(x - \bar{x})(y - \bar{y}) < 0.$
(c) $\Sigma(x - \bar{x})(y - \bar{y})$ small and near 0.
FIG. 77.—Different values of $\Sigma(x - \bar{x})(y - \bar{y})$.

There are two objections to using $\Sigma(x - \bar{x})(y - \bar{y})$ itself as a measure of association. In the first place, its magnitude depends on the units in which X and Y are measured. It would be more sensible to express deviations from the means in terms of the standard deviations s_x and s_y of X and Y respectively. Second, the magnitude of C is a function of the number of points n. A measure averaged over the n points would be better.

Our modified statistic is therefore:

$$\frac{\Sigma(x - \bar{x})(y - \bar{y})/n}{s_x s_y} = \frac{\Sigma(x - \bar{x})(y - \bar{y})}{\sqrt{\{\Sigma(x - \bar{x})^2 \, \Sigma(y - \bar{y})^2\}}}$$

$$= \frac{\Sigma xy - n\bar{x}\bar{y}}{\sqrt{\{(\Sigma x^2 - n\bar{x}^2)(\Sigma y^2 - n\bar{y}^2)\}}}. \qquad (i)$$

Although we have taken two different approaches to measuring association, it can be shown (with some rather laborious algebra) that the statistic (*i*) and the definition for $r = \sqrt{r^2}$ given in Section 7 above are entirely equivalent, *i.e.*,

$$r = \frac{(\Sigma xy - n\bar{x}\bar{y})}{\sqrt{\{(\Sigma x^2 - n\bar{x}^2)(\Sigma y^2 - n\bar{y}^2)\}}}.$$

This formula provides a computationally convenient means of calculating r, since the regression line does not have to be found in the process.

The correlation coefficient has one great advantage over the coefficient of determination in that its sign indicates whether the association is *direct* (as in Fig. 77 (*a*)) or *inverse* (as in Fig. 77 (*b*)).

9. Sampling distribution of *r*.

Unfortunately, since r^2 is calculated using sample values, it is only an estimator of the true *population coefficient of determination*, ρ^2, defined by

$$\rho^2 = 1 - \frac{\sigma_{Y|X}^2}{\sigma_Y^2}$$

where $\sigma_{Y|X}^2$ is the variance in Y-values about the regression line, and σ_Y^2 is the variance in Y. The square root of ρ^2, known as the *population correlation coefficient*, is similarly estimated by r.

The sampling distribution of r is by no means simple. However, it can be shown that if the true value of ρ is zero, then

$$\frac{r\sqrt{(n-2)}}{\sqrt{(1-r^2)}} \sim t_{n-2}.$$

This result certainly enables us to test the hypothesis that $\rho = 0$ against one- or two-sided alternative hypotheses. It does not, however, enable us to perform tests involving a null hypothesis that ρ has a non-zero value, nor to provide confidence intervals for ρ in general.

To overcome this difficulty, we use a not uncommon statistical technique. An algebraic operation is made on r in order to obtain a statistic the sampling distribution of which is known approximately. The so-called *z-transformation* appropriate here takes the following form:

$$z = \tfrac{1}{2} \log_e \left\{ \frac{1+r}{1-r} \right\}.$$

Approximately this *z*-statistic has a normal distribution with mean

$$\tfrac{1}{2} \log_e \left\{ \frac{1+\rho}{1-\rho} \right\}$$

and variance $1/(n-3)$. Its use is illustrated in the worked example.

Worked Example

Using the data in the worked example of Section 2 above, test the hypothesis that the population correlation coefficient (between total costs and output) is 0·8, using a significance level of 5%. Find also a 95% confidence interval for ρ.

Answer

(i)
$$H_0: \rho = 0.80.$$
$$H_1: \rho \neq 0.80.$$

(ii)
$$\alpha = 0.05.$$

(iii) Test statistic:

$$z = \tfrac{1}{2} \log_e \left\{ \frac{1+r}{1-r} \right\}.$$

If H_0 is true, z is normally distributed with,

$$\mu = \tfrac{1}{2} \log_e \left\{ \frac{1+0.8}{1-0.8} \right\} = \tfrac{1}{2} \log_e 9 = 1.0986$$

$$\sigma^2 = 1/(n-3) = 1/(10-3) = 0.1429.$$

Rejection region is defined by

$$\frac{z-\mu}{\sigma} > |1.96|$$

i.e.,

$$z > (1.96 \times \sqrt{0.1429}) + 1.0986 = 1.839$$

or

$$z > 1.0986 - (1.96 \times \sqrt{0.1429}) = 0.3577.$$

(iv) Sample results

$$r = \frac{679.7 - (35 \times 169.7)/10}{\sqrt{\{145 - (35 \times 35)/10\}\{3246.03 - (169.7 \times 169.7)/10\}}}$$

$$= 0.9446.$$

$$z = \tfrac{1}{2} \log_e \left(\frac{1.9446}{0.0554} \right)$$

$$= 1.7791.$$

(v) H_0 cannot be rejected.

A 95% confidence interval for the mean of the distribution of z, *i.e.* μ, can be found easily enough.

$$1.7791 - (1.96 \times 0.3780) < \mu < 1.7791 + (1.96 \times 0.3780)$$

i.e.,

$$1.04 < \mu < 2.52.$$

Now

$$\mu = \tfrac{1}{2} \log_e \left(\frac{1+\rho}{1-\rho} \right)$$

i.e.,

$$2\mu = \log_e \left(\frac{1 + \rho}{1 - \rho} \right)$$

$$\left(\frac{1 + \rho}{1 - \rho} \right) = e^{2\mu}$$

$$\therefore \quad 1 + \rho = e^{2\mu} - \rho \cdot e^{2\mu}$$

$$\rho = (e^{2\mu} - 1)/(e^{2\mu} + 1).$$

A 95% confidence interval for ρ is therefore given by:

$$(e^{2 \times 1 \cdot 04} - 1)/(e^{2 \times 1 \cdot 04} + 1) < \rho < (e^{2 \times 2 \cdot 52} - 1)/(e^{2 \times 2 \cdot 52} + 1)$$

i.e.,

$$0 \cdot 778 < \rho < 0 \cdot 987.$$

Exercises on Section 9

7. A member of the personnel department records the average wage rates of a sample of key jobs in a manufacturing plant, together with the corresponding evaluations of the jobs on a points system. His results are:

Wage rate (Y)	Job evaluation (X) (points)	X^2	Y^2	XY
4·87	8·3	68·89	23·72	40·42
4·73	10·3	106·09	22·37	48·72
4·99	12·9	166·41	24·90	64·37
5·92	12·6	158·76	35·05	74·59
6·63	14·4	207·36	43·96	95·47
6·31	16·3	265·69	39·82	102·85
6·79	15·1	228·01	46·10	102·53
7·76	18·2	331·24	60·22	141·23
8·70	17·9	320·41	75·69	155·73
8·36	25·6	655·36	69·89	214·02
8·81	23·8	566·44	77·62	209·68
9·47	16·2	262·44	89·68	153·41
Total 83·34	191·6	3337·10	609·02	1403·02

(*i*) Find the sample coefficient of determination. (*ii*) Find the sample correlation coefficient. (*iii*) Test, at the 2% significance level, the claim made by management that ρ is at least 0·90. (*iv*) Give a 96% confidence interval for ρ.

8. The sample coefficient of correlation is computed for five paired observations of sales (x_1) of a particular product and expenditure on advertising (x_2).

x_1	x_2
(£000)	(£00)
5	3
8	4
7	5
6	2
4	1

Test for existence of correlation at the 0·10 level of significance.

9. You are interested in examining the relationship between the number of customer contacts per week, and sales per week. The following data were obtained last quarter.

Salesman	Average no. of contacts/week	Total amount of sales (£000)	Average no. of sales/week
1	96	152	36
2	108	160	38
3	116	181	41
4	120	182	42
5	126	180	40
6	130	186	45
7	133	184	44
8	137	191	53
9	140	208	55
10	142	200	54

(a) Do these data indicate that the total amount of sales is associated with number of contacts?

(b) Is amount of sales associated with number of sales?

10. In Question 1, Exercises on Section 6 above, find the sample coefficient of correlation between wages and volume. Test, at the 2% significance level, the claim that $\rho < 0.90$.

11. In Question 2, Exercises on Section 6 above, calculate a 90% confidence interval for the sample correlation coefficient.

EXTENSIONS TO SIMPLE LINEAR REGRESSION ANALYSIS AND CORRELATION

10. Reduction to linear form. Because of its comparative convenience a linear model may often be assumed over a small range of the independent variable, particularly if only interpolative predictions are to be made. Indeed, it is possible to fit several line segments over different parts of the total range (*see* Fig. 78).

Furthermore, there are one or two instances where a simple transformation will reduce a more complex model to a linear form. Consider the scatter diagram of Fig. 79.

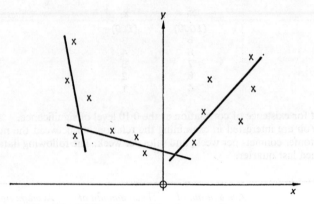

FIG. 78.—Fitting of several least squares lines segments.

A suitable model would be:

$$y = Ae^{mx}. \qquad (i)$$

By taking natural logarithms of both sides of the equation, we obtain:

$$\log_e y = mx + \log_e A.$$

Writing $y' = \log_e y$, $x' = x$, and $c = \log_e A$ this becomes:

$$y' = mx' + c. \qquad (ii)$$

The values of y' and x' can now be used to fit a regression line to (ii). The gradient and intercept of this line provide values for m, c (and therefore A).

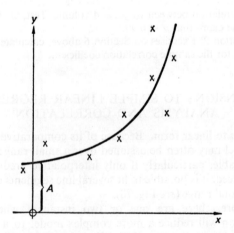

FIG. 79.—Fitting of least squares curve of form $y = Ae^{mx}$.

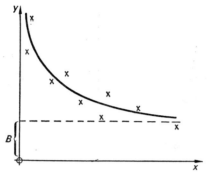

Fig. 80.—Fitting of least squares curve of form $y = A/x + B$.

In a similar way, the scatter diagram of Fig. 80 indicates a possible inverse relationship of the form:

$$y = \frac{A}{x} + B.$$

By writing $\frac{1}{x} = x'$ this transforms to a straight line

$$y = Ax' + B$$

and we can regress in the usual way to find values for A and B.

Unfortunately, transformations of this type only enable us to deal with certain special cases. The reality of the situation is that not all associations between variables are of (or reducible to) the straight line variety.

Exercises on Section 10

12. A company utilises a particularly complex machine in the grinding of crumps. A wage-incentive scheme is introduced in a number of the company's factories on a trial basis with the following results:

Factory	Wage rate (X)	% defective crumps (Y)
A	2·4	38
B	2·8	15
C	3·2	14
D	3·6	5
E	4·0	7
F	4·4	4

Suggest a suitable functional relationship between wage rate and percentage of defectives, and use this to estimate the percentage of defectives beyond which improvement is unlikely.

The scatter diagram appears in Fig. 81.

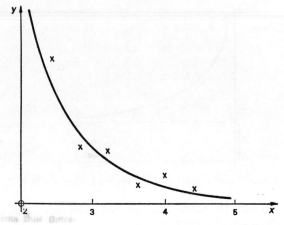

FIG. 81.—Scatter diagram for wage-rate/no. of defectives.

11. Polynomial regression. The linear function is the simplest of a series of functions known as polynomials. In ascending order of complexity we have:

$$y = a + bx + cx^2 \qquad \text{quadratic}$$
$$y = a + bx + cx^2 + dx^3 \qquad \text{cubic, etc.}$$

On the assumption that one of these functions will provide an appropriate model to our data, normal equations can be derived in exactly the same manner as in Section 2 above. The fitting of a quadratic curve, for example, in order to minimise the scatter about the curve, involves solution for a, b, c of the normal equations:

$$\Sigma y = na + b\,\Sigma x + c\,\Sigma x^2.$$
$$\Sigma xy = a\,\Sigma x + b\,\Sigma x^2 + c\,\Sigma x^3.$$
$$\Sigma x^2 y = a\,\Sigma x^2 + b\,\Sigma x^3 + c\,\Sigma x^4.$$

The normal equations for fitting polynomials of higher degree than two follow an analogous pattern. Obviously the solution of equations of this type is a tedious business. Fortunately, regressions are so frequently needed that standard computer programs are readily available.

12. Multiple regression. Particularly in the business and economic world the dependence of a variable on only one other is something of a rarity. The market price of a share, for example, is dependent on a number of factors.

By a natural extension of our previous arguments, if y is dependent on two variables, x_1 and x_2, and the model

$$y = a + bx_1 + cx_2$$

is assumed, it can be shown that the "best fit" to the data is given by using the values of a, b, and c which satisfy the normal equations:

$$\Sigma y = na + b \Sigma x_1 + c \Sigma x_2.$$
$$\Sigma x_1 y = a \Sigma x_1 + b \Sigma x_1^2 + c \Sigma x_1 x_2.$$
$$\Sigma x_2 y = a \Sigma x_2 + b \Sigma x_1 x_2 + c \Sigma x_2^2.$$

For models involving more variables, $x_1, x_2, x_3, \ldots$ the relevant normal equations follow a similar pattern.

It is impossible to explore multiple regression in detail here. The basic procedure is illustrated with a worked example. A treatment of the subject in greater depth may be found in the references in the Bibliography, p. 298.

Worked Example

An accountant of Regressive Industries Limited claims that there is a linear relationship between the unit cost of a product, y, the percentage of labour time lost through strikes, x_1, and the unit cost of raw material x_2. The model assumed is

$$y = a + bx_1 + cx_2.$$

Data are collected as follows:

$y(£)$	$x_1 y$	$x_1(\%)$	x_1^2	$x_2(p)$	$x_2 y$	x_2^2	$x_1 x_2$
65	65	1	1	23	1495	529	23
64	128	2	4	18	1152	324	36
33	99	3	9	10	330	100	30
75	300	4	16	21	1575	441	84
32	160	5	25	9	288	81	45
74	444	6	36	19	1406	361	114
Total 343	1196	21	91	100	6246	1836	332

(The additional computations have been included in the table.)
Find the regression equation.

Answer

The normal equations are:

$$343 = 6a + 21b + 100c.$$
$$1196 = 21a + 91b + 332c.$$
$$6246 = 100a + 332b + 1836c.$$

Solving these three simultaneous equations provides us with

$$a = -12.44, \ b = 3.321, \ c = 3.479.$$

Thus,

$$y = -12.44 + 3.321x_1 + 3.479x_2.$$

13. Coefficient of determination in multiple regression. For a multiple regression model, the coefficient of determination, r^2, has a similar definition and interpretation to that in the bivariate case. Using the notation of Section 7 above

$$r^2 = 1 - \frac{\Sigma(y - y_R)^2}{\Sigma(y - \bar{y})^2}$$

where y_R is the value of the dependent variable predicted by the multiple regression equation for given values of x_1 and x_2.

Take the preceding worked example by way of illustration.

y	y_R	$y - y_R$	$(y - y_R)^2$
65	−12·44 + 3·321 + 80·017 = 70·898	−5·898	34·786
64	56·824	7·176	51·495
33	32·313	0·687	0·472
75	73·903	1·097	1·203
32	35·476	−3·476	12·083
74	73·587	0·413	0·171
Total			100·21

Thus,

$$r^2 = 1 - \frac{100·21}{21,535 - 19,608} = 1 - \frac{100·21}{1926} = 0·948.$$

This value indicates that the assumed linear dependence on x_1 and x_2 explains the variation in y very well.

14. Partial coefficients of determination and correlation. Although the coefficient of determination tells us the proportion of total variation in y due to dependence on x_1 and x_2, it does not give any indication of the individual dependence on x_1, or on x_2.

Now, the proportion of the variation in y which is explained by taking account of x_1, and ignoring x_2 completely, is just the coefficient of determination, r_{y1}^2, say, between x_1 and y. The subsequent inclusion of x_2 into the model then results in the explanation of a further proportion $r^2 - r_{y1}^2$ of the total variation in y. In other words, the inclusion of x_2 enables us to explain a proportion

$$\frac{r^2 - r_{y1}^2}{1 - r_{y1}}$$

of the variation unexplained by dependence on x_1 alone. This ratio is called the *partial coefficient of determination between y and x_2*. This measure is denoted by $r_{y2\cdot1}^2$, the suffices indicating that the association between y and x_2 is being measured, whilst x_1 is taken into account but not varied.

A more convenient formula for purposes of calculation is

$$r_{y2\cdot1} = \frac{r_{y2} - r_{y1}r_{2\cdot1}}{\sqrt{(1 - r_{y1}^2)(1 - r_{2\cdot1}^2)}}$$

Here r_{y2}, r_{21} are the coefficients of determination between y and x_2, and between x_1 and x_2 respectively. In an analogous way, of course, we may define $r_{y1\cdot2}$. Again referring to the data of our worked example,

$$r_{y2\cdot1} = \frac{0\cdot9268 + (0\cdot0245)(0\cdot3306)}{\sqrt{\{1 - (0\cdot0245)^2\}}\sqrt{\{1 - (0\cdot3306)^2\}}}$$

$$= 0\cdot9909.$$

$$r_{y1\cdot2} = \frac{0\cdot0245 + (0\cdot9268)(0\cdot3306)}{\sqrt{\{1 - (0\cdot9268)^2\}}\sqrt{\{1 - (0\cdot3306)^2\}}}$$

$$= 0\cdot9337.$$

Exercises on Section 14

13. In Question 9, Exercises on Section 9, find the least squares linear regression equation of total sales (y), on number of contacts per week (x_1), and number of sales per week (x_2). Calculate the sample coefficients of determination, r^2 and the sample partial coefficients of determination $r_{y1\cdot2}^2$ and $r_{y2\cdot1}^2$.

14. Total costs associated with various output levels of completed items are as follows:

	Total costs
Output	(£)
1·1	19·8
1·8	30·8
3·3	32·4
3·6	50·6
5·5	52·2
6·0	79·2

A relationship of the form

$$y = a + bx + cx^2$$

is suspected. Find the quadratic function of "best fit" to the data.

CHAPTER 12

TIME SERIES ANALYSIS AND INDEX NUMBERS

INTRODUCTION

Often data on a particular variable are collected over a period of time —perhaps daily (*e.g.*, share prices or stock market indices of performance), monthly (*e.g.*, sales), annually (*e.g.*, net profits) or at some other regular intermediate intervals. The analysis of *time series* such as these serves two useful purposes. In the first place, the isolation of the component factors which cause variations in successive datum values provides an understanding of the basic behaviour of the variable over time. Secondly, this understanding hopefully leads in turn to an ability to predict future values of the series.

In their study of time series analysis economists and statisticians have found two models to be especially useful. Both of these assume that the values of the variable, Y, are determined by four factors:

> (*i*) The seasonal factor   (*S*).
> (*ii*) The long-term trend   (*T*).
> (*iii*) The cyclical factor   (*C*).
> (*iv*) Random effects   (*E*).

The *multiplicative model*, however, suggests that

$$Y = STCE$$

and the *additive model* that

$$Y = S + T + C + E.$$

Briefly, the seasonal factor is simply the periodic increase or decrease of the variable on a monthly, or perhaps quarterly, basis. A package tour operator, for example, will experience marked seasonal variations in his income.

As with most time series, however, his income will, over a period of years, exhibit a *long-term trend* of increase or decrease.

In addition, his income will be subject to *cyclical* upward and downward variations caused by changes in the social and political environment, and by alternate periods of economic expansion and recession.

Cycles are in general difficult to isolate since they are of indeterminate length. Finally, the operator's income will vary due to the short-term *random effects* of, for example, strikes, weather, aeroplane breakdown, and political instability. Random effects are naturally troublesome in the short term since they are by their very nature unpredictable.

The four factors are illustrated by Fig. 82 which graphs the hypothetical fortunes from 1960 to 1965 of Sunhat Tours Ltd.

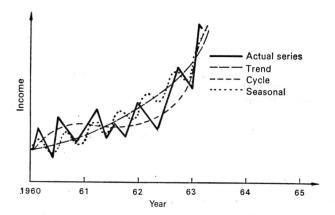

FIG. 82.—Time series of (hypothetical) tour operator's income.

We shall use the multiplicative model since it is found to explain the majority of time series more adequately than the additive model. The latter, furthermore, requires that S, C, T, and E be all measured in the units of Y, and does not make allowance for the interaction between the factors. By measuring the trend value in units of Y, and applying to it a set of adjusting ratios which express S, C, and E as percentages of trend, the multiplicative model avoids these disadvantages.

Each of the factors will now be examined in turn with reference to Table 49, which gives the raw materials cost of borborofluoride—an additive used by a manufacturer in the production of toothpaste.

The graph derived from this table is shown in Fig. 83.

LONG-TERM TREND

1. Method of moving averages. A moving average is produced by calculating over a fixed period of time—perhaps 3 months, 6 months, or a year—the average value of the variable. Each value of the series is thereby replaced by an average over a number of points in time before,

Table 49. Costs of borborofluoride

	Jan.	Feb.	Mar.	Apr.	May	Jun.	Jul.	Aug.	Sept.	Oct.	Nov.	Dec.
1965	400	414	491	450	475	491	549	549	508	447	403	458
1966	472	497	533	500	505	602	635	668	577	527	475	535
1967	541	541	651	649	632	671	729	751	654	582	497	555
1968	563	519	649	627	646	729	834	809	715	632	560	632
1969	668	643	737	742	745	869	1005	958	861	756	654	767
1970	783	765	875	864	878	1032	1140	1118	980	845	748	845
1971	869	831	983	960	980	1165	1283	1289	1115	958	842	927
1972	938	878	999	960	1002	1201	1355	1394	1115	991	856	930
1973	994	944	1121	1093	1159	1303	1512	1543	1278	1123	991	1118
1974	1151	1079	1156	1272	1303	1477	1717	1673	1402	1272	1076	1192

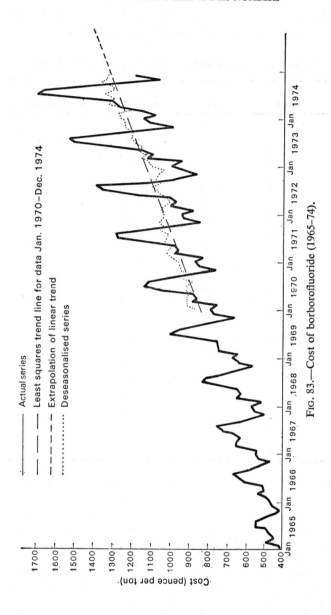

FIG. 83.—Cost of borborofluoride (1965–74).

including, and after, the particular time. Graphically the general effect is to smooth out some of the fluctuations in the series.

Table 50. Five-point moving averages using data from Table 49

Month		Cost	Five-point moving totals	Five-point moving averages
1972	J	938		
	F	878		
	M	999	→ 4777	955·4
	A	960	5040	1008·0
	M	1002	5517	1103·4
	J	1201	5912	1182·4
	J	1355	6067	1213·4
	A	1394	6056	1211·2
	S	1115	5711	1142·2
	O	991	5286	1057·2
	N	856	4886	977·2
	D	930	4715	943·0
1973	J	994	4845	969·0
	F	944	5082	1016·4
	M	1121	5311	1062·2
	A	1093	5620	1124·0
	M	1159	6188	1237·6
	J	1303	6610	1322·0
	J	1512	6795	1359·0
	A	1543	6759	1351·8
	S	1278	6447	1289·4
	O	1123	6053	1210·6
	N	991	5661	1132·2
	D	1118	5462	1092·4
1974	J	1151	5495	1099·0
	F	1079	5776	1155·2
	M	1156	5961	1192·2
	A	1272	6287	1257·4
	M	1303	6925	1385·0
	J	1477	7442	1488·4
	J	1717	7572	1514·4
	A	1673	7541	1508·2
	S	1402	7140	1428·0
	O	1272	→ 6615	1323·0
	N	1076		
	D	1192		

The procedure is illustrated in Table 50, for a five-point moving average, using the last three years' data from Table 49. Notice that the average is plotted at the centre of the interval over which it is calculated. This implies, of course, that a moving average cannot be provided for the first and last few points of the series. In this example, no values are available for initial and final two-month periods.

To present the approach as clearly as possible, a moving average over an odd number of points has been calculated. Rather more care is required in computing a moving average over, say, a six-month period. The procedure is illustrated in Table 51 for the last two years' data of Table 49.

Note that a moving average derived from the six-point totals of the second column would require plotting mid-way between months. In

Table 51. Six-point centred moving average using data from Table 49

Month		Cost	Six-point moving total	"Centred 12-point" totals	Centred six-point averages
1973	J	994			
	F	944			
	M	1121			
	A	1093	6614	13,746	1145·5
	M	1159	7132	14,863	1238·6
	J	1303	7731	15,619	1301·6
	J	1512	7888	15,806	1317·2
	A	1543	7918	15,668	1305·7
	S	1278	7750	15,315	1276·2
	O	1123	7565	14,769	1230·7
	N	991	7204	13,944	1162·0
	D	1118	6740	13,358	1113·2
1974	J	1151	6618	13,385	1115·4
	F	1079	6767	13,846	1153·8
	M	1156	7079	14,517	1209·7
	A	1272	7438	15,442	1286·8
	M	1303	8004	16,602	1383·5
	J	1477	8598	17,442	1453·5
	J	1717	8844	17,688	1474·0
	A	1673	8844	17,461	1455·1
	S	1402	8617	16,949	1412·4
	O	1272	8332		
	N	1076			
	D	1192			

order to "centre" the average, the totals for two overlapping periods of six months are summed and the result is divided by 12. Symbolically, the six-point moving average, M_t, for time t is given by:

$$M_t = \frac{\{Y_{t-3} + Y_{t-2} + Y_{t-1} + Y_t + Y_{t+1} + Y_{t+2}\}/6 + \{Y_{t-2} + Y_{t-1} + Y_t + Y_{t+1} + Y_{t+2} + Y_{t+3}\}/6}{2}$$

$$= \frac{Y_{t-3} + 2(Y_{t-2} + Y_{t-1} + Y_t + Y_{t+1} + Y_{t+2}) + Y_{t+3}}{12}$$

The unavailability of moving-average values for the first and last few points of a time series is, of course, a disadvantage from the point of view of forecasting. Additionally, the method suffers from the somewhat arbitrary device of the interval over which the average is calculated. Some skill is required in order not to use too long an interval (resulting in over-smoothing) or too short an interval (causing insufficient smoothing), although a five- or seven-point average is usually satisfactory.

Finally, if the series happens to possess a systematic fluctuation, the period of which is the same as that chosen for the moving average, then the former is removed completely. Despite its drawbacks, however, the moving average is relatively simple to calculate, and does have an important use in *deseasonalising* series (*see* Section 4 below).

2. Least squares trend line. A more sophisticated approach to the location of the trend rests on the use of regression analysis. The analyst may, from a visual inspection of the data, suspect the existence of, say, a linear, a quadratic, or an exponential trend. This is followed by the fitting of an appropriate least squares regression line (or curve) of the variable on time. Examination of Fig. 83 suggests that a linear trend of costs over time might well exist. The calculations involved in finding the "line of best fit" are considerably simplified if time, X, is measured from zero at the midpoint of the series. Our normal equations:

$$\Sigma Y = nc + m\,\Sigma X$$
$$\Sigma Y = c\,\Sigma X + m\,\Sigma X^2$$

then reduce to

$$\Sigma Y = nc$$
$$\Sigma XY = m\,\Sigma X^2$$

since the evenly-spaced positive and negative values of X on either side of $X = 0$ cancel out on summation.

Table 52. Regression analysis of data from Table 49 using coded times over 60-month period

Month		Time (X) (coded)	Cost (Y)	XY	X²
1970	J	−29·5	783	−23,098·5	870·25
	F	−28·5	765	−21,802·5	812·25
	M	−27·5	875	−24,062·5	756·25
	A	−26·5	864	−22,896·0	702·25
	M	−25·5	878	−22,389·0	650·25
	J	−24·5	1032	−25,284·0	600·25
	•	•	•	•	•
	•	•	•	•	•
	•	•	•	•	•
1972	J	−5·5	938	−5159·0	30·25
	F	−4·5	878	−3951·0	20·25
	M	−3·5	999	−3496·5	12·25
	A	−2·5	960	−2400·0	6·25
	M	−1·5	1002	−1503·0	2·25
	J	−0·5	1201	−600·5	0·25
	J	+0·5	1355	+677·5	0·25
	A	+1·5	1394	+2091·0	2·25
	S	+2·5	1115	+2787·5	6·25
	O	+3·5	991	+3468·5	12·25
	N	+4·5	856	+3852·0	20·25
	D	+5·5	930	+5115·0	30·25
	•	•	•	•	•
	•	•	•	•	•
	•	•	•	•	•
1974	J	+24·5	1717	+42,066·5	600·25
	A	+25·5	1673	+42,661·5	650·25
	S	+26·5	1402	+37,153·0	702·25
	O	+27·5	1272	+34,980·0	756·25
	N	+28·5	1076	+30,666·0	812·25
	D	+29·5	1192	+35,164·0	870·25
Total		$\Sigma X = 0$	$\Sigma Y = 65{,}643$	$\Sigma XY = 147{,}849{\cdot}5$	$\Sigma X^2 = 17{,}995{\cdot}0$

(Similar advantages of cancellation are obtained when fitting a higher degree curve if the time values are coded in this way.)

The full details are shown in Table 52 and Fig. 83 using the data of Table 49.

Since $\Sigma Y = 65,643$, $\Sigma XY = 147,849 \cdot 5$, $\Sigma X^2 = 17,995$, and $n = 60$, the normal equations give:

$$m = \Sigma XY / \Sigma X^2 = 147,849 \cdot 5 / 17,995 = 8 \cdot 2161$$

and

$$c = \Sigma Y / n = 1094 \cdot 05.$$

The least squares regression line is therefore:

$$Y = 8 \cdot 22X + 1094 \cdot 05.$$

It is now a simple matter to extrapolate this line over a few time periods in order to obtain forecasts of future trend values.

For purposes of simplicity, the least squares method has been exemplified in the case of a suspected linear trend. Over a relatively short period of time the assumption of linearity is in general a reasonable one. However, over the long term the trend may take a different form. Some of the possibilities are illustrated in Fig. 84.

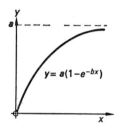

(a) Sales of pocket calculators.

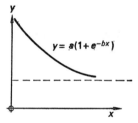

(b) Sales of slide rules.

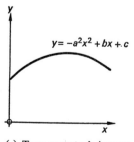

(c) Tour operator's income from "package" holidays.

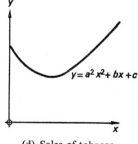

(d) Sales of tobacco.

FIG. 84.—Some typical trend lines (x = time).

Exercises on Section 2

1. The following table gives the sales of electrical energy in millions of kilowatt hours of Ruritania from 1953 to 1974:

Year	Sales/kWh	Year	Sales/Kwh
1953	3·2	1964	5·6
1954	3·3	1965	6·2
1955	3·4	1966	6·4
1956	3·5	1967	6·8
1957	3·6	1968	7·2
1958	4·1	1969	7·9
1959	4·2	1970	8·3
1960	4·6	1971	8·8
1961	4·9	1972	9·2
1962	5·0	1973	9·9
1963	5·5	1974	10·3

Use a seven-point moving average to find the trend line.

2. The annual imports of steel products of Lilliput from 1965 to 1974 are as follows (thousand tons):

157·8	117·8	136·8	140·2	130·2
139·2	149·2	167·8	183·4	178·0

Find the least squares trend line.

THE SEASONAL FACTOR

3. Calculation of seasonal indices. In the following discussion "seasonal" is used in its more or less literal sense. However, the term is applicable to any short-term systematic variations superimposed on the trend. Depending on the particular time series under scrutiny it may be quite sensible, to talk of "seasonal" effects of, say, a day's or a week's duration.

The technique primarily used for isolating the seasonal component of a time series is the so-called *ratio to moving average*. In our example we shall assume that systematic variations in costs are expected to be on a monthly basis, and commence our argument with the calculation of a twelve-point moving average for the series. The purpose of this is twofold. In the first place, by taking a one-year interval in which each month is equally represented, the smoothing effect will hopefully remove seasonal effects from the data.

Secondly, by smoothing over a period of this length the random component of the series should be largely eliminated.

In theory, then, a twelve-month moving average will contain only trend and cycle components, and division of the time series values by this average will leave the seasonal and random components only. Algebraically, we have,

$$Y = STCE.$$

$$\frac{Y}{TC} = \frac{STCE}{TC} = SE.$$

The analysis is shown in Tables 53 and 54 using the data of Table 49. That it is sensible to talk of seasonal indices is substantiated by the

Table 53. Calculation of ratio to centred twelve-point moving average

Month		Cost (Y)	12-point totals	"24"-point totals	Centred 12-point moving average	Y/moving average	Deseasonalised data
1970	J	783					865·3
	F	765					905·5
	M	875					902·2
	A	864					903·0
	M	878					889·9
	J	1032	10,873				909·7
	J	1140	10,959	21,832	909·7	1·2532	900·9
	A	1118	11,025	21,984	916·0	1·2205	879·0
	S	980	11,133	22,158	923·2	1·0615	924·5
	O	845	11,229	22,362	931·7	0·9069	916·9
	N	748	11,331	22,560	940·0	0·7957	934·1
	D	845	11,464	22,795	949·8	0·8897	944·8
1971	J	869	11,607	23,071	961·3	0·9040	960·3
	F	831	11,778	23,385	974·4	0·8528	983·7
	M	983		·	·	·	1013·5
	·	·	·	·	·	·	·
	·	·	·	·	·	·	·
	·	·	·	·	·	·	·
1973	O	1123	14,685	·	·	·	1218·5
	N	991	14,829	29,514	1229·7	0·8059	1237·5
	D	1118	15,003	29,832	1243·0	0·8994	1250·0
1974	J	1151	15,208	30,211	1258·8	0·9144	1272·0
	F	1079	15,338	30,546	1272·7	0·8478	1277·2
	M	1156	15,462	30,800	1283·3	0·9008	1191·9
	A	1272	15,611	31,073	1294·7	0·9825	1329·4
	M	1303	15,696	31,307	1304·5	0·9989	1320·7
	J	1477	15,770	31,466	1311·1	1·1265	1302·0
	J	1717					1356·9
	A	1673					1315·4
	S	1402					1322·6
	O	1272					1380·2
	N	1076					1343·7
	D	1192					1332·7

similar appearance of the graphs in Fig. 85. Here the ratio to moving average is plotted month-by-month for each year in turn.

The figures appearing in the penultimate column of Table 53 are estimates of the composite factor *SE*, and as such do not provide genuine seasonal indices. An obvious way to eliminate *E* is to average in turn

Table 54. Complete listing of ratio to moving average
(as calculated in Table 53)

Year	Month					
	J	F	M	A	M	J
1970	—	—	—	—	—	—
1971	0·9040	0·8528	0·9958	0·9625	0·9742	1·1496
1972	0·9057	0·8418	0·9538	0·9153	0·9536	1·1423
1973	0·8946	0·8399	0·9860	0·9511	0·9989	1·1102
1974	0·9144	0·8478	0·9008	0·9825	0·9989	1·1265

Year	Month					
	J	A	S	O	N	D
1970	1·2532	1·2205	1·0615	0·9069	0·7957	0·8897
1971	1·2582	1·2582	1·0855	0·9321	0·8185	0·8990
1972	1·2857	1·3163	1·0452	0·9198	0·7857	0·8452
1973	1·2726	1·2855	1·0585	0·9233	0·8059	0·8994
1974	—	—	—	—	—	—

the four estimates which are available for each month. Since means
are affected by outlying values, however, a preferable alternative is to
form the median of each set of estimates. A second alternative (which
gives the same results in the particular case when four estimates are
available) is to ignore the least and greatest estimate, and form the
average of those remaining.

For convenience the penultimate column of Table 53 is rearranged
in Table 54, with derived seasonal indices provided in Table 55.

Table 55. Seasonal indices (calculated as the median of the four-values
of ratio to moving average given in Table 54)

Month	Seasonal Index
January	0·9049
February	0·8448
March	0·9699
April	0·9568
May	0·9866
June	1·1344
July	1·2654
August	1·2719
September	1·0600
October	0·9216
November	0·8008
December	0·8944

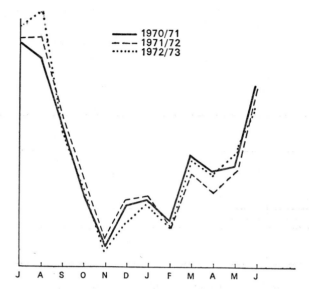

FIG. 85.—Ratio to moving average (plotted month by month, 1970/71, 1971/72, 1972/73).

Seasonal indices are useful both analytically and synthetically, since they enable the effects of seasonal factors on the variable to be determined, and they facilitate prediction by superimposing a seasonal effect on an extrapolated trend line.

With the data of Table 52, and the trend line of Section 2 above, trend forecasts for the next three months are:

Jan. 1974: ($t = 30\cdot5$): $Y_t = 8\cdot22 (30\cdot5) + 1094\cdot05 = 1344\cdot76$.
Feb. 1974: ($t = 31\cdot5$): $Y_t = 8\cdot22 (31\cdot5) + 1094\cdot05 = 1352\cdot98$.
Mar. 1974: ($t = 32\cdot5$): $Y_t = 8\cdot22 (32\cdot5) + 1094\cdot05 = 1361\cdot20$.

Adjusting these forecasts for seasonal effects then gives:

Jan. 1974: $Y_t = 1344\cdot76 \times 0\cdot9049 = 1216\cdot9$.
Feb. 1974: $Y_t = 1352\cdot98 \times 0\cdot8448 = 1143\cdot0$.
Mar. 1974: $Y_t = 1361\cdot20 \times 0\cdot9699 = 1320\cdot2$.

This approach is naturally only of value in the production of short-term forecasts, since the longer-term cyclical component is not incorporated in the synthesis.

4. Deseasonalised data. It is sometimes useful in evaluating the significance of recent points in a time series to be able to *deseasonalise* the data, *i.e.* to remove the "smoke-screen" effect of seasonal variations.

To accomplish this, the actual values of the series are divided by the appropriate seasonal index. The final column of Table 53 gives deseasonalised values using the indices of Table 55 as divisors.

The results are shown graphically in Fig. 83.

Exercises on Section 4

3.

			Month			
Year	J	F	M	A	M	J
1971	6·25	5·75	6·95	6·90	7·00	6·45
1972	6·50	6·30	7·10	6·95	7·10	6·55
1973	6·80	6·15	7·05	6·90	7·15	6·65
1974	6·65	6·00	7·00	6·90	7·10	6·65

			Month			
Year	J	A	S	O	N	D
1971	6·35	6·15	5·85	6·15	6·10	6·40
1972	6·50	6·30	6·05	6·35	6·25	6·65
1973	6·65	6·40	6·15	6·45	6·30	6·60
1974	6·60	6·50	6·35	6·65	6·70	7·10

The data above give the exports of crude oil (million barrels) from Bahman. Find the least squares trend line, and calculate the seasonal indices.

4. Deseasonalise the time series of Question 3.

5.

		Year		
Month	1971	1972	1973	1974
J	1477	1499	1529	1570
F	1492	1507	1545	1580
M	1505	1525	1557	1596
A	1517	1547	1586	1615
M	1543	1563	1609	1640
J	1561	1583	1620	1662
J	1561	1592	1629	1681
A	1566	1588	1625	1672
S	1555	1577	1607	1642
O	1561	1586	1613	1659
N	1542	1575	1610	1652
D	1531	1559	1599	1652

The data above give the average monthly wages (in cents) of the workers at a Ruritanian distillery. Find the least squares trend line and the seasonal indices.

6. Deseasonalise the series of Question 5.

THE CYCLICAL FACTOR

5. Cyclical indices. If the time series data are given on an annual basis, then seasonal variations are not present, and:

$$Y = TCE.$$

By estimating a least squares trend line the trend values, T, can be obtained and used to eliminate this component from the series. Thus,

$$\frac{Y}{T} = CE.$$

Monthly data can be deseasonalised after fitting the trend line, and the resulting values divided by trend values to arrive at a similar composite measure of the cyclical and random effects.

Although it would be feasible to aggregate the values of a time series of monthly data into annual totals in order to remove seasonal effects, it is not advisable to do so. By using monthly data it is possible to determine a cycle more precisely, and in particular to find its turning points with an accuracy of approximately one month, rather than of a year. Even so, from a forecasting viewpoint cycles are troublesome, since they are generally of indeterminate length.

Having isolated CE we are left to take the final step of isolating the cycle. The simplest and most often used technique is to form a moving average of the CE indices with the objective of smoothing out the random component. In practice a moving average of four or five months is usually found to be reasonably satisfactory. Averaging over a longer period runs the risk of smoothing away the cyclical component itself, whilst over a shorter period may fail to remove the random effects adequately.

The detailed results of the analysis for our original time series are given in Table 56. The cyclical variations are sometimes expressed as *percentages of trend, i.e.*

$$\frac{Y}{TS} \times 100,$$

or alternatively as *cyclical residuals, i.e.*

$$\frac{Y - TS}{TS} \times 100 = \left(\frac{Y}{TS} - 1\right) 100.$$

Exercises on Section 5

7. Using the data and results of Questions 3 and 4 in Exercises on Section 4 above, calculate the cyclical-random components, the cyclical indices, and the cyclical residuals.

Table 56. Calculation of Cyclical Index

Year	Month	Cost (Y)	Trend (T)	Seasonal Index (S)	$CE = \dfrac{Y}{TS}$	Cycle C (centred four-point MA)
1970	J	1140	901	1·2654	0·9999	
	A	1118	909	1·2719	0·9670	
	S	980	917	1·0600	1·0082	0·9913
	O	845	926	0·9216	0·9902	0·9959
	N	748	934	0·8008	1·0001	1·0007
	D	845	942	0·8944	1·0029	1·0056
1971	J	869	950	0·9049	1·0109	1·0162
	F	831	958	0·8448	1·0268	·
	M	983	967	0·9699	1·0481	·
	·	·	·	·	·	·
	·	·	·	·	·	·
	·	·	·	·	·	·
1973	O	1123	1221	0·9216	0·9980	·
	N	991	1230	0·8008	1·0061	·
	D	1118	1238	0·8944	1·0097	1·0112
1974	J	1151	1246	0·9049	1·0208	1·0060
	F	1079	1254	0·8448	1·0185	1·0027
	M	1156	1263	0·9699	0·9437	1·0087
	A	1272	1271	0·9568	1·0460	1·0094
	M	1303	1279	0·9866	1·0326	
	J	1477	1287	1·1344	1·0117	

8. Using the data and results of Questions 5 and 6 in Exercises on Section 4 above, calculate the cyclical-random components, the cyclical indices, and the cyclical residuals.

INDEX NUMBERS

6. Introduction. Although index numbers do not strictly appear under the "umbrella" of time series analysis, their construction usually does arise from the consideration of changing prices or quantities over time, and it is therefore convenient to introduce them at this stage.

The basic purpose in calculating an index number (or, more concisely, an *index*) is to relate the values of a number of variables in a given period of time to the values of the same variables in a different period. The latter is known as the *base* period. Some of the more important types of indices, together with their interpretations, are described in the next few sections. The following notation will be used throughout:

p_0 = price of a commodity in base period.
p_n = price of a commodity in period n.
q_0 = quantity of a commodity in base period.
q_n = quantity of a commodity in period n.

7. Aggregative indices. The data of Table 57 show the prices per ton of borborofluoride from May to October 1974.

Table 57. Simple price relatives of borborofluoride
(May 1974 = 100)

Month	Price/ton	Simple price relative (%)
1974 May	1303	100·0
June	1477	113·4
July	1717	131·8
August	1673	128·4
September	1402	107·6
October	1272	97·6

In the final column are given the *simple price relatives*. Following normal practice, the results are expressed as percentages of the price at the base date, which is taken here to be March 1974.

Although this index might be of value to the toothpaste manufacturer, it is more common to calculate composite price indices which simultaneously compare prices of a number of commodities over a period of time. Take, for example, the data given in Table 58 on the prices in cents of several foods in Lilliput.

Table 58. (Unweighted) Aggregative Price Index (1971 = 100)

Year	Milk (per pint)	Bread (per 2lb loaf)	Beef (per lb)	Σ (prices)	Aggregative Index (%)
1971	2·4	7·0	43·0	52·4	100·0
1972	3·3	7·6	55·6	66·5	126·9
1973	4·0	10·8	72·1	86·9	165·8
1974	5·5	12·5	80·3	98·3	187·6

Using a base-date of 1971, the *aggregative index* calculated in the final column is defined by:

$$\frac{\Sigma p_n}{\Sigma p_0} \times 100\%.$$

There are two objections to this index. In the first place the use of different units of measurement can have a distorting effect. Suppose that instead of the price of milk being given per pint, it was given per gallon. The aggregative indices would then have been as in Table 59.

Table 59. (Unweighted) Aggregative Price Index (1971 = 100): effect of measuring price of milk in gallons rather than pints

Year	Milk (per gall)	Bread (per 2lb loaf)	Beef (per lb)	Σ (prices)	Aggregative Index (%)
1971	19·2	7·0	43·0	69·2	100·0
1972	26·4	7·6	55·6	89·6	129·5
1973	32·0	10·8	72·1	114·9	166·0
1974	44·0	12·5	80·3	136·8	197·7

Secondly, if this type of composite index is intended to be used as a measure of the rise in cost of living, then account should also be taken of the quantities of each commodity which are consumed. Accordingly, it is advisable to calculate a *weighted aggregative index*, in which weights are attached to the prices in proportion to the importance of the commodities.

The *Laspeyres Index*, defined by

$$\frac{\Sigma p_n q_0}{\Sigma p_0 q_0} \times 100\%$$

is of this type. Its calculation is illustrated in Table 60. The data used are the same as in Table 58, with the additional information concerning the average weekly per capita consumption of each commodity in the base-period.

Paasche's Index is of similar conception to the Laspeyres Index, but uses as weights the quantities, q_n, relevant to the current period rather than the base period. In symbols it is defined by:

$$\frac{\Sigma p_n q_n}{\Sigma p_0 q_n} \times 100\%.$$

It is interesting to compare these two indices. The Laspeyres Index assumes that from the base period onwards the quantities of the commodities consumed remain the same. Paasche's Index has the advantage that changing patterns of consumer purchasing behaviour are taken into account, although of course it may be expensive and time-consuming to revise the weights for every time period. Although Paasche's Index is more realistic, it is rather inconvenient from a computational viewpoint, due to the fact that the denominator is different for each period.

Table 60. Calculation of Laspeyres and Paasche Indices (1971 = 100)

	1971 p_0	1971 q_0	1972 p_1	1972 q_1	1973 p_2	1973 q_2	1974 p_3	1974 q_3	p_0q_0	p_1q_0	p_2q_0	p_3q_0	p_0q_1	p_0q_2	p_0q_3	p_1q_1	p_2q_2	p_3q_3
Milk	2·4	10	3·3	8	4·0	15	5·5	20	24·0	33·0	40·0	55·0	19·2	36·0	48·0	26·4	60·0	110·0
Bread	7·0	4	7·6	8	10·8	2	12·5	3	28·0	30·4	43·2	50·0	56·0	14·0	21·0	60·8	21·6	37·5
Beef	43·0	3	55·6	2	72·1	7	80·3	5	129·0	166·8	216·3	240·9	86·0	301·0	215·0	111·2	504·7	401·5
Total									181·0	230·2	299·5	345·9	161·2	351·0	284·0	198·4	586·3	549·0

(a)

Year	Laspeyres Index $\{\Sigma p_n q_0 / \Sigma p_0 q_0\} \times 100$	Paasche Index $\{\Sigma p_n q_n / \Sigma p_0 q_n\} \times 100$
1971 ($n = 0$)	100·0	100·0
1972 ($n = 1$)	127·2	123·1
1973 ($n = 2$)	165·5	167·0
1974 ($n = 3$)	191·1	193·3

(b)

Exercises on Section 7

9. The following table gives the unit prices of five components used in the manufacture of stocks:

Year	Hooks	Lines	Sinkers	Locks	Barrels
1970	25·3	13·8	8·1	6·4	3·1
1971	29·2	14·1	9·3	7·3	3·7
1972	37·4	15·2	8·9	7·9	4·0
1973	42·7	17·3	10·1	8·4	4·4
1974	48·1	17·8	11·2	8·8	5·1

(*i*) Calculate the unweighted aggregative index.

(*ii*) If in 1970 a complete stock required 3 hooks, 2 lines, 4 sinkers, 4 locks, and 1 barrel calculate the appropriate Laspeyres Price Index from 1970 to 1974 (1971 value = 100).

(*iii*) Due to technological advances and the redesigning of stocks, the 1974 model requires 2 hooks, 1 line, 5 sinkers, 3 locks, and 2 barrels. Calculate the appropriate Paasche Price Index.

8. Average of price relatives. The aggregative indices of Section 7 above do not provide the only means of constructing a composite index concerning several commodities. A simple (*unweighted*) *average of price relatives* is defined by:

$$\frac{\Sigma\left(\dfrac{p_n}{p_0}\right) \times 100}{N}$$

where N is the number of commodities under consideration. Its calculation is illustrated (with reference to the earlier data of Table 58) in Table 61.

Table 61. Average Price Relatives (1971 = 100)

Year	Price relatives, $p_n/p_0 \times 100$			Average Price Relatives
	Milk	Bread	Beef	
1971 (*n* = 0)	100·0	100·0	100·0	100·0
1972 (*n* = 1)	137·5	108·6	129·3	125·1
1973 (*n* = 2)	166·7	154·3	167·7	162·9
1974 (*n* = 3)	229·2	178·6	186·7	198·2

Although this index is not affected by differences in the units in which prices are measured, it does have the disadvantage of attaching equal importance to each commodity.

To remedy this a *weighted average of price relatives* can be used instead:

$$\frac{\sum p_0 q_0 \left(\dfrac{p_n}{p_0}\right) 100}{\sum p_0 q_0}.$$

The weights used here are the *values* (*i.e.*, price × quantity) of each commodity in the base-period. It would not be appropriate to use either prices or quantities alone for weighting purposes, since the weighted relatives would not then be all in the same units.

Known as the *Laspeyres average of price relatives*, the above index has a natural analogue—the *Paasche average of price relatives*, defined by:

$$\frac{\sum p_0 q_n \left(\dfrac{p_n}{p_0}\right) 100}{\sum p_0 q_n}.$$

In fact, although reached by different routes, these indices are easily seen to be algebraically identical to their counterparts of Section 7 above.

9. Changing the base of an index. In general it is not difficult to change the base-period of an index number. The procedure is illustrated in Table 62 with reference to the Laspeyres Index calculated in Section 7 above.

Table 62. Change of base-period

Year	Laspeyres Index (1971 = 100)	Laspeyres Index (1973 = 100)
1971	100·0	$\dfrac{100 \times 100}{165\cdot5} = 60\cdot4$
1972	127·2	$\dfrac{127\cdot2 \times 100}{165\cdot5} = 76\cdot9$
1973	165·5	$\dfrac{165\cdot5 \times 100}{165\cdot5} = 100\cdot0$
1974	191·1	$\dfrac{191\cdot1 \times 100}{165\cdot5} = 115\cdot5$

Exercises on Section 9

10. Using the data of Question 9, Exercises on Section 7.
   (*i*) Calculate the unweighted average price relatives.
   (*ii*) Calculate the Laspeyres and Paasche averages of price relatives.
   (*iii*) Change the base-period of the Laspeyres indices in (*ii*) from 1970 to 1973.

10. Concluding comments. Although our work has concentrated on price indices, for each of these there does in fact exist an analogous quantity index, used for comparing consumption or output from one period to another.

We have, for instance, a Laspeyres Index of Quantities:

$$\frac{\Sigma p_0 q_n}{\Sigma p_0 q_0} \times 100.$$

In general, the formula for a quantity index can be obtained from the corresponding price index formula by simply interchanging p and q.

For detailed discussion of certain special indices, such as the index of retail prices, the interested reader may consult the references in the Bibliography, p. 298.

CHAPTER 13

THE ELEMENTS OF STATISTICAL DECISION THEORY

DECISIONS AND DECISION CRITERIA

1. Definitions. Recent years have seen the emergence of an increasingly important approach to decision-making, which uses the concepts of probability and statistical inference. Known as *statistical (or Bayesian) decision theory* this approach can prove an invaluable aid in the analysis of complex modern business decisions.

In order to provide a framework in which to introduce the subject, it is useful to distinguish clearly the features common to all situations involving decision-making. The *decision-maker* (who may be an individual, a group of individuals, or perhaps a company) is faced with the selection of one from a set of possible *courses of action (or acts)* open to him. In addition there is a mutually exclusive and exhaustive set of *events* (or *states of nature*) which describe the possible behaviours of the environment in which the decision is made. The decision-maker has no control over which event will take place, and can only attach a subjective probability (or *degree of belief*) of occurrence to each. Finally, with each combination of a course of action and an event is associated a *pay-off*, which measures the net benefit to the decision-maker of that particular combination.

To illustrate these definitions, consider the company whose board of directors is presented with three different investment programmes, each concerned with the development of the same new product with a view to its appearance on the market in four years' time. Because of its novelty the eventual sales of the product are uncertain, but are divided for convenience into four categories.

The detailed analysis of the decision situation is shown in a pay-off table (Table 63).

The pay-offs vary between investment programmes since each involves a different initial outlay, and different "spin-off" which will help in the manufacture of other products. We will assume that pay-offs are based on the first two years' sales. The figures in brackets will be used in Section 3 below.

2. The maximin decision criterion. This simple criterion for choice between alternative course of action assumes a pessimistic view of nature. Taking each act in turn, we note the worst possible resulting

Table 63. Pay-off table for investment decision

Events	Course of Action		
	Adopt investment programme A	Adopt investment programme B	Adopt investment programme C
Very bad sales (0·10)	−16,000	−12,000	−4000
Fair sales (0·25)	−6000	−2000	−500
Good sales (0·45)	6000	3000	1500
Excellent sales (0·20)	15,000	16,000	5000

pay-off, and select the act which maximises this minimum pay-off. Referring to Table 63:

Course of action	Worst pay-off
A	−16,000
B	−12,000
C	−4000

Act *C* is selected since it has a greater minimum pay-off than either of the other two.

Although this criterion has the merit of simplicity, if a businessman were to use it consistently the unrealistic assumption of pessimism would prevent him from developing his company. In addition, the method takes no account of the differing probabilities of events.

3. Maximum expected pay-off. Usually, either from subjective assessment, sample information, or a combination of both, probabilities of occurrence can be attached to the various events. If this can be done and the pay-offs carefully evaluated, the criterion of *maximum expected pay-off* (otherwise known as *Bayes decision rule*) proves to be optimal.

The bracketed figures in Table 63 give the probabilities of the different pay-offs. Since the events are mutually exclusive and exhaustive the probabilities must, of course, sum to one. For each course of action the expected pay-off can be calculated:

Programme A
$$\text{Expected pay-off} = \{(-16{,}000 \times 0{\cdot}10) + (-6000 \times 0{\cdot}25) + (6000 \times 0{\cdot}45) + (15{,}000 \times 0{\cdot}20)\}$$
$$= \text{£}2600.$$

Programme B
$$\text{Expected pay-off} = \{(-12{,}000 \times 0{\cdot}10) + (-2000 \times 0{\cdot}25) + (3000 \times 0{\cdot}45) + (16{,}000 \times 0{\cdot}20)\}$$
$$= \text{£}2850.$$

Programme C
$$\text{Expected pay-off} = \{(-4000 \times 0\cdot10) + (-500 \times 0\cdot25) \\ + (1500 \times 0\cdot45) + (5000 \times 0\cdot20)\} \\ = £1150.$$

Since it has the maximum expected pay-off, programme B is chosen.

4. Opportunity loss. An opportunity of realising a greater pay-off is foregone if the best course of action is not taken. The difference between this greater pay-off, and the actual pay-off is known as the *opportunity loss*. In accounting and finance we meet the more specific concept of an *opportunity cost*.

With reference again to the data of Table 63 we see that if the event "fair sales" actually occurs the best course of action would have been "adopt programme *C*." The opportunity loss of having instead taken the course of action "adopt programme *A*" is therefore

$$£\{-500 - (-6000)\} = £5500.$$

Proceeding in this way leads to an opportunity loss table (Table 64).

Table 64. Opportunity loss table for investment decision

Events	Course of Action		
	Adopt investment programme A	Adopt investment programme B	Adopt investment programme C
Very bad sales (0·10)	12,000	8000	0
Fair sales (0·25)	5500	1500	0
Good sales (0·45)	0	3000	4500
Excellent sales (0·20)	1000	0	11,000

Using the bracketed probabilities in this Table the *expected opportunity loss* (EOL) corresponding to each course of action can be found:

Adopt programme A
$$\text{EOL} = £\{(12,000 \times 0\cdot10) + (5500 \times 0\cdot25) + (0 \times 0\cdot45) \\ + (1000 \times 0\cdot20)\} \\ = £2775.$$

Adopt programme B
$$\text{EOL} = £\{(8000 \times 0\cdot10) + (1500 \times 0\cdot25) + (3000 \times 0\cdot45) \\ + (0 \times 0\cdot20)\} \\ = £2525.$$

Adopt programme C

$$\text{EOL} = \pounds\{(0 \times 0.10) + (0 \times 0.25) + (4500 \times 0.45)$$
$$+ (11{,}000 \times 0.20)\}$$
$$= \pounds 4225.$$

Under the decision criterion of minimising expected opportunity loss, act *B* is chosen. Note that this choice is the same as that made under the criterion of Section 3 above. It can be shown that this is the case in general.

5. Expected value of perfect information. Before the decision to take a particular course of action has been made the existence of a perfectly reliable forecast of which state of nature will occur would clearly be of value to the decision-maker.

Suppose that such a predictor was known to exist, but had not yet been used to supply the prediction of which state of nature will occur. In the context of our investment decision, our decision-maker will receive a pay-off of $-\pounds 4000$ with a probability of 0.10, of $-\pounds 500$ with a probability of 0.25, of $\pounds 6000$ with a probability of 0.45, and of $\pounds 16{,}000$ with a probability of 0.20, since whichever event is eventually predicted he will choose the course of action which will maximise his pay-off. Accordingly, his *expected pay-off with perfect information* (EPPI) is given by:

$$\text{EPPI} = \{(-4000 \times 0.10) + (-500 \times 0.25) + (6000 \times 0.45)$$
$$+ (16{,}000 \times 0.20)\}$$
$$= \{-400 - 125 + 2700 + 3200\}$$
$$= \pounds 5375.$$

If the decision were made under uncertainty (*i.e.*, in the absence of a source of perfect information), however, the maximum expected pay-off is $\pounds 2850$, resulting from the adoption of programme *B*. Accordingly the expected value to be placed on perfect information (EVPI) is seen to be:

$$\text{EVPI} = \text{EPPI} - \text{maximum expected pay-off under uncertainty}$$
$$= \pounds 5375 - \pounds 2850$$
$$= \pounds 2525.$$

It is interesting to note that the EVPI (or *cost of uncertainty*) is equal to the previously calculated minimum expected opportunity loss. This can be shown to be true in general.

Exercises on Section 5

1. A wholesale newsagent and bookseller is preparing his orders. In particular, he assigns the following probability distribution to the sales of a specialist annual publication:

Number sold	Probability
0	0·08
1	0·10
2	0·20
3	0·35
4	0·18
5	0·09

Each volume costs the newsagent £13, and sells for £26. Any unsold copies can be returned to the publisher who will refund £6·50 per copy. Every volume he sells costs the newsagent £1·30 in packaging, transaction, and delivery costs.

(*i*) What is the newsagent's optimal ordering policy?

(*ii*) What is the expected value of perfect information?

2. Given the pay-off table below:

(*i*) Find the optimal act, using the criterion of maximising expected pay-off.

(*ii*) Find the expected value of perfect information.

(*iii*) Find the minimum expected opportunity loss.

| Events | Probability | I | Courses of action | | |
			II	III	IV
E_1	1/8	−10	14	20	20
E_2	1/2	44	30	−10	26
E_3	1/4	36	−24	16	10
E_4	1/8	20	16	24	−14

BAYESIAN DECISION ANALYSIS

6. Prior and posterior analysis. In Sections 1–5 above, the decision-making process has been analysed on the basis of prior beliefs—expressed by *prior probabilities*—concerning the occurrence of events.

However, the methods of sampling which we have discussed earlier at some length can be used to provide additional information about future (or unknown) events. This information can then be used to revise the *prior probabilities* to form *posterior probabilities*. In other words, the results of sampling can be used to up-date our beliefs. Schematically:

Prior beliefs (probabilities) + sample information →
Posterior beliefs (probabilities)

7. The total probability theorem and Bayes theorem. In order to use additional information properly to derive posterior probabilities, two important theorems are needed. Although at first sight these may appear rather fearsome, the worked example should assist in providing a clear understanding of both their interpretation and application. Since in formulating the theorems use will be made of set theoretic concepts, it may be of value to reread Chapter 2, Sections 1–3, before proceeding.

Suppose we have an event A contained in a sample space Ω, and that the latter is subdivided into three mutually exclusive and exhaustive events H_1, H_2, H_3 (see Fig. 86). In fact any number of such events

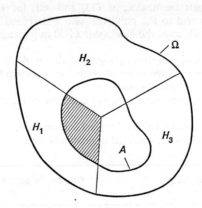

FIG. 86.—Illustration of total probability theorem.

can be defined—three are taken simply for definiteness. Typically this subdivision of Ω will simultaneously divide A into three events, of which the shaded area is an example. In fact, since this latter event is common to both A and H_1 it is $A \cap H_1$, and the others are $A \cap H_2$, $A \cap H_3$, respectively. Since these three events are exhaustive (as far as A is concerned) and mutually exclusive:

$$P(A) = P(A \cap H_1) + P(A \cap H_2) + P(A \cap H_3) \qquad (i)$$

(see Chapter 2, Section 14).

Now, for any two events the conditional probability of one given the other is defined by

$$P(A \mid H_1) = \frac{P(A \cap H_1)}{P(H_1)}$$

(see Chapter 3, Sections 3–5).

Therefore,

$$P(A \cap H_1) = P(A \mid H_1)P(H_1). \qquad (ii)$$

Replacing each term of (*i*) by the corresponding expression (*ii*) leads to.

$$P(A) = P(A \mid H_1)P(H_1) + P(A \mid H_2)P(H_2) + P(A \mid H_3)P(H_3). \quad (iii)$$

This equation is known as the *total probability theorem*. Now, for any two sets,

$$P(H_1 \mid A) = \frac{P(H_1 \cap A)}{P(A)} = \frac{P(A \cap H_1)}{P(A)}$$

$$= \frac{P(A \mid H_1)P(H_1)}{P(A)}$$

from (*ii*).

Using (*iii*) we finally have:

$$P(H_1 \mid A) = \frac{P(A \mid H_1)P(H_1)}{P(A \mid H_1)P(H_1) + P(A \mid H_2)P(H_2) + P(A \mid H_3)P(H_3)}. \quad (iv)$$

The last equation is *Bayes Theorem*.

Worked Example

An investor's degree of belief that a share will increase in value over the next week is 0·90. Dow Trusts, an investment advisory firm, then predicts that the share will not do so. In the past the firm has predicted no increase correctly 75% of the time, and has predicted no increase in 15% of the cases where an increase did take place. What is the investor's revised degree of belief that the share will increase in value?

Answer

Define the following events:

$H_1 \equiv$ {share increases in value}.
$H_2 \equiv$ {share does not increase in value}.
$A \equiv$ {Dow Trusts predict that share will not increase in value}.

We seek $P(H_1 \mid A)$.

Noting that H_1, H_2 are mutually exclusive and exhaustive events, we may apply (*iv*) above:

$$P(H_1 \mid A) = \frac{P(A \mid H_1)P(H_1)}{P(A \mid H_1)P(H_1) + P(A \mid H_2)P(H_2)}$$

Using the information provided,

$$P(A \mid H_1) = 0·15.$$
$$P(A \mid H_2) = 0·75.$$

Initially,

$$P(H_1) = 0·90$$

and

$$P(H_2) = 0·10.$$

Thus,

$$P(H_1|A) = \frac{0.15 \times 0.90}{(0.15 \times 0.90) + (0.75 \times 0.10)}$$

$$= \frac{0.135}{0.21} = 0.64.$$

As we would expect, the original high degree of belief in a future increase has been reduced by the reasonably reliable contrary prediction made by the firm.

It is worth drawing attention to the common misconception that $P(A|H_1)$ and $P(A|H_2)$ should total to unity. Given that H_1 happens, the events A and $\bar{A}$ are complementary, *i.e.*

$$P(A|H_1) + P(\bar{A}|H_1) = 1.$$

Similarly,

$$P(A|H_2) + P(\bar{A}|H_2) = 1.$$

The events "A, given H_1" and "A, given H_2" are not, however, complementary.

Incidentally, if we required $P(H_2|A)$, this would be found from:

$$P(H_2|A) = \frac{P(A|H_2)P(H_2)}{P(A|H_1)P(H_1) + P(A|H_2)P(H_2)}$$

$$= \frac{0.75 \times 0.10}{(0.15 \times 0.90) + (0.75 \times 0.10)} = 0.36.$$

Alternatively, $H_1|A$ and $H_2|A$ are complementary since, given that A happens then either H_1 or H_2 (and not both) happens. Accordingly,

$$P(H_2|A) = 1 - P(H_1|A)$$
$$= 0.36.$$

Finally, it is instructive to look at the problem under conditions of (*i*) perfectly reliable and (*ii*) perfectly unreliable information from Dow Trusts.

In case (*i*),

$$P(A|H_1) = 0$$
$$P(A|H_2) = 1$$

and

$$P(H_1|A) = \frac{0 \times 0.90}{(0 \times 0.90) + (1 \times 0.10)} = 0$$

$$P(H_2|A) = 1.$$

In other words given a (perfectly reliable) prediction of increase from Dow Trusts, our investor's belief in increase rises to 1, and in decrease falls to zero. This is so whatever his prior beliefs were.

In case (*ii*),

$$P(A|H_1) = 1$$
$$P(A|H_2) = 0$$

and

$$P(H_1|A) = \frac{1 \times 0.90}{(1 \times 0.90) + (0 \times 0.10)} = 1$$

$$P(H_2|A) = 0.$$

In the typical situation of partially reliable information, the posterior beliefs lie between the extreme values of (i) and (ii).

Exercises on Section 7

3. The Cantech Exploration Co. is involved in a mining operation in Canada. Their chief geologist originally feels that there is a 50–50 chance of finding an economic mineral deposit. A test drill subsequently gives favourable results. From previous experience it is known that the probability of obtaining misleading results (either way) is 0·3. What is the geologist's revised belief in the existence of an economic deposit?

4. A firm is considering marketing one of its products in a litre container, rather than the current $1\frac{1}{2}$-litre size. Their marketing manager considers the probability of increasing profits as a result of this change to be 0·75. In a limited test area the change is found to cause reduced profits. The reliability of the test is such that this result would occur with a probability of 0·3, even if a change nationally would increase profits, and with a probability of 0·8 if a change nationally would reduce profits. What is the manager's revised belief in the profitability of a change?

8. Posterior expected pay-offs under uncertainty. The expected pay-offs calculated in Section 3 above were dependent on a set of prior probabilities for the possible states of nature. Let us now suppose that before making its investment decision, the company commissions a market research organisation to investigate the marketability of the intended product. The results of this sample survey of potential customers provide the company with information which can be used to derive a set of posterior probabilities for the four events. Finally, a revised set of *posterior expected pay-offs* can be found and used to select the appropriate course of action. Since the rankings of posterior and prior expected pay-offs may not be the same, the effect of the survey results may be to alter the decision.

For the sake of argument we assume the survey to indicate that sales will be good, and we denote this result by G. Referring to the events "very bad sales," "fair sales," "good sales," and "excellent sales" as H_1, H_2, H_3, and H_4 respectively, the reliability of the survey is summarised by:

$$P(G|H_1) = 0.10 \qquad P(G|H_3) = 0.60$$
$$P(G|H_2) = 0.12 \qquad P(G|H_4) = 0.15$$

From Table 63 we also have:

$$P(H_1) = 0.10 \qquad P(H_3) = 0.45$$
$$P(H_2) = 0.25 \qquad P(H_4) = 0.20$$

Thus,

$$P(G) = \sum_{i=1}^{4} P(G|H_i)P(H_i)$$

(from the Total Probability Theorem)

$$= \{(0\cdot10 \times 0\cdot10) + (0\cdot12 \times 0\cdot25) + (0\cdot60 \times 0\cdot45)$$
$$+ (0\cdot15 \times 0\cdot20)\}$$
$$= 0\cdot34.$$

Therefore,

$$P(H_1|G) = \frac{P(G|H_1)P(H_1)}{P(G)} = \frac{0\cdot10 \times 0\cdot10}{0\cdot34} = 0\cdot0294.$$

$$P(H_2|G) = \frac{0\cdot12 \times 0\cdot25}{0\cdot34} = 0\cdot0882.$$

$$P(H_3|G) = \frac{0\cdot60 \times 0\cdot45}{0\cdot34} = 0\cdot7941.$$

$$P(H_4|G) = \frac{0\cdot15 \times 0\cdot20}{0\cdot34} = 0\cdot0882.$$

These, then, are the *posterior probabilities* of the events, derived from the prior probabilities and survey information.

Our posterior expected pay-offs are seen to be:

Programme A
  Posterior Expected pay-off

$$= £\{(0\cdot0294 \times -16{,}000) + (0\cdot0882 \times -6000)$$
$$+ (0\cdot7941 \times 6000) + (0\cdot0882 \times 15{,}000)\}$$
$$= £5088.$$

Programme B
  Posterior Expected pay-off

$$= £\{(0\cdot0294 \times -12{,}000) + (0\cdot0882 \times -2000)$$
$$+ (0\cdot7941 \times 3000) + (0\cdot0882 \times 16{,}000)\}$$
$$= £3264\cdot3.$$

Programme C
  Posterior Expected pay-off

$$= £\{(0\cdot0294 \times -4000) + (0\cdot0882 \times -500)$$
$$+ (0\cdot7941 \times 1500) + (0\cdot0882 \times 5000)\}$$
$$= £1470\cdot45.$$

Using now the decision criterion of maximising the expected pay-off, programme A is chosen. Notice that the survey information has changed the previous decision of Section 3, which was to adopt programme B.

9. Expected value of sample information.
In the previous section the result of sampling was to provide information which increased the (prior) expected pay-off under the optimal act (adopt programme B) from £2850 to a value of £5088 (*i.e.*, the posterior expected pay-off under the optimal act, *viz.* adopt A). By a simple extension of this line of argument we can incorporate into our model the decision to sample or not. Since the following takes place before this decision is made, it is referred to as *pre-posterior analysis*.

To proceed, we need further information concerning the reliability of our suggested survey. This is provided by the market research

Table 65. Conditional probabilities associated with market research organisation's report

Events	$P(H_i)$	$P(B|H_i)$	$P(F|H_i)$	$P(G|H_i)$	$P(E|H_i)$
H_1	0·10	0·65	0·23	0·10	0·02
H_2	0·25	0·20	0·62	0·12	0·06
H_3	0·45	0·08	0·12	0·60	0·20
H_4	0·20	0·04	0·06	0·15	0·75

organisation before entering into agreement with them, and is summarised in Table 65. The notation used is the same as that in Section 8 above, with the additions:

$B \equiv$ {survey indicates bad sales}.
$F \equiv$ {survey indicates fair sales}.
$E \equiv$ {survey indicates excellent sales}.

The figures given in the fourth column have, of course, already been used in Section 8 above.

From Section 8,

$$P(G) = \sum_{i=1}^{4} P(G|H_i)P(H_i)$$
$$= 0·34.$$

Similarly,

$$P(B) = \sum_{i=1}^{4} P(B|H_i)P(H_i)$$
$$= (0·65 \times 0·10) + (0·20 \times 0·25)$$
$$+ (0·08 \times 0·45) + (0·04 \times 0·20)$$
$$= 0·159.$$

$$P(F) = (0.23 \times 0.10) + (0.62 \times 0.25)$$
$$+ (0.12 \times 0.45) + (0.06 \times 0.20)$$
$$= 0.244.$$
$$P(E) = (0.02 \times 0.1) + (0.06 \times 0.25)$$
$$+ (0.2 \times 0.45) + (0.75 \times 0.20)$$
$$= 0.257.$$

(Note that $P(G) + P(B) + P(F) + P(E) = 1$, since B, G, F, E form a mutually exclusive and exhaustive set of outcomes for the survey.)

We can now derive the complete set of posterior probabilities shown in Table 66.

Table 66. Posterior event probabilities

Events	$P(H_i|B)$	$P(H_i|F)$	$P(H_i|G)$	$P(H_i|E)$
H_1	0.4088	0.0943	0.0294	0.0078
H_2	0.3147	0.6352	0.0882	0.0584
H_3	0.2264	0.2213	0.7941	0.3502
H_4	0.0503	0.0492	0.0882	0.5837

The posterior expected pay-offs associated with each possible sample result and course of action are calculated as in Section 8, and shown in full in Table 67.

Table 67. Posterior expected pay-offs

Survey results	A	Acts B	C
B	−6316	−4051	−1201
F	−3254	− 951	− 117
G	5088	3264	1470
E	10,382	10,179	3383

On the basis of these figures, we can now argue that *if* the survey is commissioned, and *if* its prediction is "bad sales" then act C would be chosen since it has the maximum posterior expected pay-off of −£1201. The survey will give such a result with a probability $P(B) = 0.159$. In a similar manner, a survey would predict "fair sales" with a probability $P(F) = 0.244$, leading to a posterior expected pay-off of −£117, since act C would again be chosen as optimal.

In summary, the commissioning of a survey will give the following posterior expected pay-offs:

Posterior expected pay-off	Probability
−1201	0·159
− 117	0·244
5088	0·340
10,382	0·257

Finally, we find the expected pay-off, if the survey is commissioned, and the optimal action taken on the basis of the sample information, i.e.

$$£\{(-1201 \times 0·159) + (-117 \times 0·244) + (5088 \times 0·340)$$
$$+ (10,382 \times 0·257)\}$$
$$= £(-190·96 - 28·55 + 1729·92 + 2668·17)$$
$$= £4178·58.$$

Now if no sample were taken, our maximum expected pay-off (using only prior probabilities) is £2850 (see Section 3 above). If a survey is commissioned the expected value is increased to that above. Accordingly the *expected value of sample information* (EVSI), is:

$$EVSI = £(4178·58 - 2850)$$
$$= £1328·58.$$

The market research organisation will, of course, charge for their services. Thus, the *expected net value of sample information* (ENSI) is:

$$ENSI = EVSI - \text{cost of survey}.$$

Naturally a positive value for ENSI would be followed by the decision to carry out the survey, whilst a negative value would argue for basing the investment decision on prior probabilities alone.

Exercises on Section 9

5. An impresario is considering the launching of a new play. The play may be either filmed and sold to a television company, or it may be produced on stage. The pay-off table is as follows:

	Produced on stage	Sold to T.V. company
Play a success	15,000	9000
Play a failure	−6000	−3000

The impresario assesses the prior probability of a success at 0·35, but is considering the possibility of having a number of critics read the play before he goes ahead with production. In the past the critics have given unfavourable reports on 72% of the plays which subsequently were a failure, and favourable reports on 80% of plays which were a success eventually.

(*i*) What is the prior expected pay-off for each act? Which is optimal?
(*ii*) What is the posterior expected pay-off for each act?
(*iii*) Find the expected value of perfect information.
(*iv*) What is the expected value of the sample (*i.e.*, critics') information?
(*v*) Suppose the critics' report is favourable. What is then the optimal act?

6. A finance company is considering investing in a new business enterprise. The possible outcomes are that the business will be successful (*S*), partially successful (*P*), or unsuccessful (*U*), with prior probabilities 0·60, 0·25, and 0·15 respectively.

The company's opportunity loss table is given below (in £00,000):

Event	Invest	Do not invest
S	0	3
P	2	0
U	5	0

An investigation into the possible viability of the business is proposed, the results of which will indicate either success (*S*), partial success (*P*), or lack of success (*U*).

The posterior probabilities of *S*, *P*, or *U*, given the investigation conclusions, are:

		Events		
		S	P	U
Results	S	0·75	0·18	0·07
of	P	0·20	0·65	0·15
investigation	U	0·10	0·18	0·72

(*e.g.*, $P(S|P) = 0·20$, etc.)

Provide the finance company with an analysis of the situation.

BIBLIOGRAPHY

The following suggestions for further reading include alternative treatments of various topics, more advanced work in particular areas, and specific applications of some of the techniques that have been discussed.

General

Chacko, G. K.	*Applied Statistics in Decision Making*. Elsevier, New York, 1971.
Chou, Ya-Lun	*Statistical Analysis*. Holt, Rinehart and Winston, New York, 1969.
Clark, C. T. and Schkade, L. L.	*Statistical Methods for Business Decisions*. South-Western Publishing Company, Cincinnati, 1969.
Dyckman, T. R., Smidt, S. and McAdams, A. K.	*Management Decision-Making under Uncertainty*. Macmillan, London, 1969.
Hamburg, M.	*Statistical Analysis for Decision-Making*. Harcourt, Brace and World, New York, 1970.
Kim, C.	*Statistical Analysis for Induction and Decision*. Dryden Press, Illinois, 1973.
Lapin, L. L.	*Statistics for Modern Business Decisions*. Harcourt, Brace, Jovanovich, New York, 1973.
Larson, H. J.	*Introduction to Probability Theory and Statistical Inference*. John Wiley and Sons, Inc., New York, 1969.
Moore, P. G.	*Risk in Business Decision*. Longman, London, 1972.
Sasaki, K.	*Statistics for Modern Business Decision Making*. Wadsworth, California, 1968.

Of particular relevance to:

Chapter 7

Hansen, M. H., Hurwitz, W. N. and Madow, W. G.	*Sample Survey Methods and Theory*. John Wiley and Sons, Inc., New York, 1965.
Iriji, Y. and Kaplan, R. S.	*A Model for Integrating Sampling Objectives in Auditing. Journal of Accounting Research*, Vol. 9, 1971, pp. 73–87. (Reprinted in Dickinson, J. P. (ed.) *Risk and Uncertainty in Accounting and Finance*. D. C. Heath Ltd., Farnborough, 1974.)

McRae, T. W.   *Statistical Sampling for Audit and Control.* John
                              Wiley and Sons Ltd., London, 1974.

Chapter 9

Kempthorne, O.   *The Design and Analysis of Experiments.* John
                              Wiley and Sons, Inc., New York, 1952.

Chapter 10

Siegel, S.   *Non-Parametric Statistics for the Behavioural
                              Sciences.* McGraw-Hill, New York, 1956 (in-
                              cludes tables for use with the Wilcoxon test and
                              Wald-Wolfowitz Runs test).

Chapter 11

Koutsoyiannis, A.   *Theory of Econometrics.* Macmillan, London, 1973
                              (examines regression in depth).
Thomas, J. J.   *An Introduction to Statistical Analysis for Econo-
                              mists.* Weidenfeld and Nicolson, London, 1973
                              (includes derivation of normal equations).

Chapter 12

Nelson, C. R.   *Applied Time Series Analysis for Managerial Fore-
                              casting.* Holden-Day, San Francisco, 1973.
Robinson, C.   *Business Forecasting: an economic approach.* Nel-
                              son, London, 1971.
Wheelwright, S. C. and   *Forecasting Methods for Management.* John Wiley
Makridakis, S.   and Sons, Inc., New York, 1973.

Chapter 13

Raiffa, H. and   *Applied Statistical Decision Theory.* Harvard
Schaifer, R.   University, Boston, 1961.

Advanced work drawing heavily on statistical theory:

Dickinson, J. P. (ed.)   *Portfolio Analysis.* D. C. Heath Ltd., Farnborough,
                              1974.
Dickinson, J. P. (ed.)   *Risk and Uncertainty in Accounting and Finance.*
                              D. C. Heath Ltd., Farnborough, 1974.
Fama, E. F. and   *The Theory of Finance.* Holt, Rinehart and Wins-
Miller, M. H.   ton, New York, 1972.
Francis, J. C. and   *Portfolio Analysis.* Prentice-Hall, N. J., 1971.
Archer, S. H.
Haley, C. W. and   *The Theory of Financial Decisions.* McGraw-Hill,
Schall, L. D.   New York, 1973.
Jean, W. H.   *The Analytical Theory of Finance.* Holt, Rinehart
                              and Winston, New York, 1972.

Mao, J. C. T.   *Quantitative Analysis of Financial Decisions.* Macmillan, New York, 1969.

Markowitz, H. M.   *Portfolio Selection: Efficient Diversification of Investments.* John Wiley and Sons, Inc., New York, 1959.

Nelson, C. R.   *The Term Structure of Interest Rates.* Basic Books, New York, 1972.

Solnik, B. H.   *European Capital Markets.* Lexington Books, Mass., 1973.

Szego, G. P. and   *Mathematical Methods in Investment and Finance.*
Shell, K.   North Holland-American Elsevier, New York, 1972.

TABLE A

Areas of the Standardised Normal Probability Distribution

The table gives the shaded area, e.g., $\Phi(1\cdot53) = 0\cdot4370 + 0\cdot5000$
$= 0\cdot9370.$

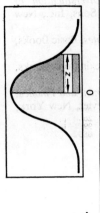

$z = \dfrac{x-\mu}{\sigma}$	0·00	0·01	0·02	0·03	0·04	0·05	0·06	0·07	0·08	0·09
0·0	0·0000	0·0040	0·0080	0·0120	0·0160	0·0199	0·0239	0·0279	0·0319	0·0359
0·1	0·0398	0·0438	0·0478	0·0517	0·0557	0·0596	0·0636	0·0675	0·0714	0·0753
0·2	0·0793	0·0832	0·0871	0·0910	0·0948	0·0987	0·1026	0·1064	0·1103	0·1141
0·3	0·1179	0·1217	0·1255	0·1293	0·1331	0·1368	0·1406	0·1443	0·1480	0·1517
0·4	0·1554	0·1591	0·1628	0·1664	0·1700	0·1736	0·1772	0·1808	0·1844	0·1879
0·5	0·1915	0·1950	0·1985	0·2019	0·2054	0·2088	0·2123	0·2157	0·2190	0·2224
0·6	0·2257	0·2291	0·2324	0·2357	0·2389	0·2422	0·2454	0·2486	0·2518	0·2549
0·7	0·2580	0·2612	0·2642	0·2673	0·2704	0·2734	0·2764	0·2794	0·2823	0·2852
0·8	0·2881	0·2910	0·2939	0·2967	0·2995	0·3023	0·3051	0·3078	0·3106	0·3133
0·9	0·3159	0·3186	0·3212	0·3238	0·3264	0·3289	0·3315	0·3340	0·3365	0·3389
1·0	0·3413	0·3438	0·3461	0·3485	0·3508	0·3531	0·3554	0·3577	0·3599	0·3621
1·1	0·3643	0·3665	0·3686	0·3708	0·3729	0·3749	0·3770	0·3790	0·3810	0·3830
1·2	0·3849	0·3869	0·3888	0·3907	0·3925	0·3944	0·3962	0·3980	0·3997	0·4015
1·3	0·4032	0·4049	0·4066	0·4082	0·4099	0·4115	0·4131	0·4147	0·4162	0·4177
1·4	0·4192	0·4207	0·4222	0·4236	0·4251	0·4265	0·4279	0·4292	0·4306	0·4319
1·5	0·4332	0·4345	0·4357	0·4370	0·4382	0·4394	0·4406	0·4418	0·4429	0·4441
1·6	0·4452	0·4463	0·4474	0·4484	0·4495	0·4505	0·4515	0·4525	0·4535	0·4545
1·7	0·4554	0·4564	0·4573	0·4582	0·4591	0·4599	0·4608	0·4616	0·4625	0·4633
1·8	0·4641	0·4649	0·4656	0·4664	0·4671	0·4678	0·4686	0·4693	0·4699	0·4706
1·9	0·4713	0·4719	0·4726	0·4732	0·4738	0·4744	0·4750	0·4756	0·4761	0·4767
2·0	0·4772	0·4778	0·4783	0·4788	0·4793	0·4798	0·4803	0·4808	0·4812	0·4817
2·1	0·4821	0·4826	0·4830	0·4834	0·4838	0·4842	0·4846	0·4850	0·4854	0·4857
2·2	0·4861	0·4864	0·4868	0·4871	0·4875	0·4878	0·4881	0·4884	0·4887	0·4890
2·3	0·4893	0·4896	0·4898	0·4901	0·4904	0·4906	0·4909	0·4911	0·4913	0·4916
2·4	0·4918	0·4920	0·4922	0·4925	0·4927	0·4929	0·4931	0·4932	0·4934	0·4936
2·5	0·4938	0·4940	0·4941	0·4943	0·4945	0·4946	0·4948	0·4949	0·4951	0·4952
2·6	0·4953	0·4955	0·4956	0·4957	0·4959	0·4960	0·4961	0·4962	0·4963	0·4964
2·7	0·4965	0·4966	0·4967	0·4968	0·4969	0·4970	0·4971	0·4972	0·4973	0·4974
2·8	0·4974	0·4975	0·4976	0·4977	0·4977	0·4978	0·4979	0·4979	0·4980	0·4981
2·9	0·4981	0·4982	0·4982	0·4983	0·4984	0·4984	0·4985	0·4985	0·4986	0·4986
3·0	0·49865	0·4987	0·4987	0·4988	0·4988	0·4989	0·4989	0·4989	0·4990	0·4990
3·1	0·49903	0·4991	0·4991	0·4991	0·4992	0·4992	0·4992	0·4992	0·4990	0·4993

TABLE B

Random Sampling Numbers

20 17	42 28	23 17	59 66	38 61	02 10	86 10	51 55	92 52	44 25
74 49	04 49	03 04	10 33	53 70	11 54	48 63	94 60	94 49	57 38
94 70	49 31	38 67	23 42	29 65	40 88	78 71	37 18	48 64	06 57
22 15	78 15	69 84	32 52	32 54	15 12	54 02	01 37	38 37	12 93
93 29	12 18	27 30	30 55	91 87	50 57	58 51	49 36	12 53	96 40
45 04	77 97	36 14	99 45	52 95	69 85	03 83	51 87	85 56	22 37
44 91	99 49	89 39	94 60	48 49	06 77	64 72	59 26	08 51	25 57
16 23	91 02	19 96	47 59	89 65	27 84	30 92	63 37	26 24	23 66
04 50	65 04	65 65	82 42	70 51	55 04	61 47	88 83	99 34	82 37
32 70	17 72	03 61	66 26	24 71	22 77	88 33	17 78	08 92	73 49
03 64	59 07	42 95	81 39	06 41	20 81	92 34	51 90	39 08	21 42
62 49	00 90	67 86	93 48	31 83	19 07	67 68	49 03	27 47	52 03
61 00	95 86	98 36	14 03	48 88	51 07	33 40	06 86	33 76	68 57
89 03	90 49	28 74	21 04	09 96	60 45	22 03	52 80	01 79	33 81
01 72	33 85	52 40	60 07	06 71	89 27	14 29	55 24	85 79	31 96
27 56	49 79	34 34	32 22	60 53	91 17	33 26	44 70	93 14	99 70
49 05	74 48	10 55	35 25	24 28	20 22	35 66	66 34	26 35	91 23
49 74	37 25	97 26	33 94	42 23	01 28	59 58	92 69	03 66	73 82
20 26	22 43	88 08	19 85	08 12	47 65	65 63	56 07	97 85	56 79
48 87	77 96	43 39	76 93	08 79	22 18	54 55	93 75	97 26	90 77
08 72	87 46	75 73	00 11	27 07	05 20	30 85	22 21	04 67	19 13
95 97	98 62	17 27	31 42	64 71	46 22	32 75	19 32	20 99	94 85
37 99	57 31	70 40	46 55	46 12	24 32	36 74	69 20	72 10	95 93
05 79	58 37	85 33	75 18	88 71	23 44	54 28	00 48	96 23	66 45
55 85	63 42	00 79	91 22	29 01	41 39	51 40	36 65	26 11	78 32
67 28	96 25	68 36	24 72	03 85	49 24	05 69	64 86	08 19	91 21
85 86	94 78	32 59	51 82	86 43	73 84	45 60	89 57	06 87	08 15
40 10	60 09	05 88	78 44	63 13	58 25	37 11	18 47	75 62	52 21
94 55	89 48	90 80	77 80	26 89	87 44	23 74	66 20	20 19	26 52
11 63	77 77	23 20	33 62	62 19	29 03	94 15	56 37	14 09	47 16
64 00	26 04	54 55	38 57	94 62	68 40	26 04	24 25	03 61	01 20
50 94	13 23	78 41	60 58	10 60	88 46	30 21	45 98	70 96	36 89
66 98	37 96	44 13	45 05	34 59	75 85	48 97	27 19	17 85	48 51
66 91	42 83	60 77	90 91	60 90	79 62	57 66	72 28	08 70	96 03
33 58	12 18	02 07	19 40	21 29	39 45	90 42	58 84	85 43	95 67
52 49	40 16	72 40	73 05	50 90	02 04	98 24	05 30	27 25	20 88
74 98	93 99	78 30	79 47	96 92	45 58	40 37	89 76	84 41	74 68
50 26	54 30	01 88	69 57	54 45	69 88	23 21	05 69	93 44	05 32
49 46	61 89	33 79	96 84	28 34	19 35	28 73	39 59	56 34	97 07
19 65	13 44	78 39	73 88	62 03	36 00	25 96	86 76	67 90	21 68
64 17	47 67	87 59	81 40	72 61	14 00	28 28	55 86	23 38	16 15
18 43	97 37	68 97	56 56	57 95	01 88	11 89	48 07	42 60	11 92
65 58	60 87	51 09	96 61	15 53	66 81	66 88	44 75	37 01	28 88
79 90	31 00	91 14	85 65	31 75	43 15	45 93	64 78	34 53	88 02
07 23	00 15	59 05	16 09	94 42	20 40	63 76	65 67	34 11	94 10
90 08	14 24	01 51	95 46	30 32	33 19	00 14	19 28	40 51	92 69
53 82	62 02	21 82	34 13	41 03	12 85	65 30	00 97	56 30	15 48
98 17	26 15	04 50	76 25	20 33	54 84	39 31	23 33	59 64	96 27
08 91	12 44	82 40	30 62	45 50	64 54	65 17	89 25	59 44	99 95
37 21	46 77	84 77	67 39	85 54	97 37	33 41	11 74	90 50	29 62

TABLE C
t Distribution

Degrees of freedom	$t_{0.100}$	$t_{0.050}$	$t_{0.025}$	$t_{0.010}$	$t_{0.005}$
1	3·078	6·314	12·706	31·821	63·657
2	1·886	2·920	4·303	6·965	9·925
3	1·638	2·353	3·182	4·541	5·841
4	1·533	2·132	2·776	3·747	4·604
5	1·476	2·015	2·571	3·365	4·032
6	1·440	1·943	2·447	3·143	3·707
7	1·415	1·895	2·365	2·998	3·499
8	1·397	1·860	2·306	2·896	3·355
9	1·383	1·833	2·262	2·821	3·250
10	1·372	1·812	2·228	2·764	3·169
11	1·363	1·796	2·201	2·718	3·106
12	1·356	1·782	2·179	2·681	3·055
13	1·350	1·771	2·160	2·650	3·012
14	1·345	1·761	2·145	2·624	2·977
15	1·341	1·753	2·131	2·602	2·947
16	1·337	1·746	2·120	2·583	2·921
17	1·333	1·740	2·110	2·567	2·898
18	1·330	1·734	2·101	2·552	2·878
19	1·328	1·729	2·093	2·539	2·861
20	1·325	1·725	2·086	2·528	2·845
21	1·323	1·721	2·080	2·518	2·831
22	1·321	1·717	2·074	2·508	2·819
23	1·319	1·714	2·069	2·500	2·807
24	1·318	1·711	2·064	2·492	2·797
25	1·316	1·708	2·060	2·485	2·787
26	1·315	1·706	2·056	2·479	2·779
27	1·314	1·703	2·052	2·473	2·771
28	1·313	1·701	2·048	2·467	2·763
29	1·311	1·699	2·045	2·462	2·756
infinity	1·282	1·645	1·960	2·326	2·576

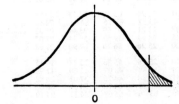

The tabulated values are those which the t-statistic exceeds with the indicated probabilities, e.g.,

$$\text{Prob.} \, (t_{10} \geqslant 1·812) = 0·05.$$

TABLE D
Chi-Square Distribution

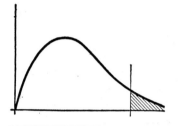

$$e.g., Prob. (\chi_{10}^2 \geqslant 20\cdot48) = 0\cdot025.$$

Degrees of freedom	Probability that chi-square value will be exceeded									
	0·995	0·990	0·975	0·950	0·900	0·100	0·050	0·025	0·010	0·005
1	0·04393	0·03157	0·03982	0·02393	0·0158	2·71	3·84	5·02	6·63	7·88
2	0·0100	0·0201	0·0506	0·103	0·211	4·61	5·99	7·38	9·21	10·60
3	0·072	0·115	0·216	0·352	0·584	6·25	7·81	9·35	11·34	12·84
4	0·207	0·297	0·484	0·711	1·064	7·78	9·49	11·14	13·28	14·86
5	0·412	0·554	0·831	1·145	1·61	9·24	11·07	12·83	15·09	16·75
6	0·676	0·872	1·24	1·64	2·20	10·64	12·59	14·45	16·81	18·55
7	0·989	1·24	1·69	2·17	2·83	12·02	14·07	16·01	18·48	20·28
8	1·34	1·65	2·18	2·73	3·49	13·36	15·51	17·53	20·09	21·96
9	1·73	2·09	2·70	3·33	4·17	14·68	16·92	19·02	21·67	23·59
10	2·16	2·56	3·25	3·94	4·87	15·99	18·31	20·48	23·21	25·19
11	2·60	3·05	3·82	4·57	5·58	17·28	19·68	21·92	24·72	26·76
12	3·07	3·57	4·40	5·23	6·30	18·55	21·03	23·34	26·22	28·30
13	3·57	4·11	5·01	5·89	7·04	19·81	22·36	24·74	27·69	29·82
14	4·07	4·66	5·63	6·57	7·79	21·06	23·68	26·12	29·14	31·32
15	4·60	5·23	6·26	7·26	8·55	22·31	25·00	27·49	30·58	32·80
16	5·14	5·81	6·91	7·96	9·31	23·54	26·30	28·85	32·00	34·27
17	5·70	6·41	7·56	8·67	10·09	24·77	27·59	30·19	33·41	35·72
18	6·26	7·01	8·23	9·39	10·86	25·99	28·87	31·53	34·81	37·16
19	6·84	7·63	8·91	10·12	11·65	27·20	30·14	32·85	36·19	38·58
20	7·43	8·26	9·59	10·85	12·44	28·41	31·41	34·17	37·57	40·00
21	8·03	8·90	10·28	11·59	13·24	29·62	32·67	35·48	38·93	41·40
22	8·64	9·54	10·98	12·34	14·04	30·81	33·92	36·78	40·29	42·80
23	9·26	10·20	11·69	13·09	14·85	32·01	35·17	38·08	41·64	44·18
24	9·89	10·86	12·40	13·85	15·66	33·20	36·42	39·36	42·98	45·56
25	10·52	11·52	13·12	14·61	16·47	34·38	37·65	40·65	44·31	46·93
26	11·16	12·20	13·84	15·38	17·29	35·56	38·89	41·92	45·64	48·29
27	11·81	12·88	14·57	16·15	18·11	36·74	40·11	43·19	46·96	49·64
28	12·46	13·56	15·31	16·93	18·94	37·92	41·34	44·46	48·28	50·99
29	13·12	14·26	16·05	17·71	19·77	39·09	42·56	45·72	49·59	52·34
30	13·79	14·95	16·79	18·49	20·60	40·26	43·77	46·98	50·89	53·67
40	20·71	22·16	24·43	26·51	29·05	51·80	55·76	59·34	63·69	66·77
50	27·99	29·71	32·36	34·76	37·69	63·17	67·50	71·42	76·15	79·49
60	35·53	37·48	40·48	43·19	46·46	74·40	79·08	83·30	88·38	91·95
70	43·28	45·44	48·76	51·74	55·33	85·53	90·53	95·02	100·4	104·22
80	51·17	53·54	57·15	60·39	64·28	96·58	101·9	106·6	112·3	116·32
90	59·20	61·75	65·65	69·13	73·29	107·6	113·1	118·1	124·1	128·3
100	67·33	70·06	74·22	77·93	82·36	118·5	124·3	129·6	135·8	140·2

F-Distribution—Upper 5% points

v_2 \ v_1	1	2	3	4	5	6	7	8	9	10	12	15	20	24	30	40	60	120	∞
1	161.4	199.5	215.7	224.6	230.2	234.0	236.8	238.9	240.5	241.9	243.9	245.9	248.0	249.1	250.1	251.1	252.2	253.3	254.3
2	18.51	19.00	19.16	19.25	19.30	19.33	19.35	19.37	19.38	19.40	19.41	19.43	19.45	19.45	19.46	19.47	19.48	19.49	19.50
3	10.13	9.55	9.28	9.12	9.01	8.94	8.89	8.85	8.81	8.79	8.74	8.70	8.66	8.64	8.62	8.59	8.57	8.55	8.53
4	7.71	6.94	6.59	6.39	6.26	6.16	6.09	6.04	6.00	5.96	5.91	5.86	5.80	5.77	5.75	5.72	5.69	5.66	5.63
5	6.61	5.79	5.41	5.19	5.05	4.95	4.88	4.82	4.77	4.74	4.68	4.62	4.56	4.53	4.50	4.46	4.43	4.40	4.36
6	5.99	5.14	4.76	4.53	4.39	4.28	4.21	4.15	4.10	4.06	4.00	3.94	3.87	3.84	3.81	3.77	3.74	3.70	3.67
7	5.59	4.74	4.35	4.12	3.97	3.87	3.79	3.73	3.68	3.64	3.57	3.51	3.44	3.41	3.38	3.34	3.30	3.27	3.23
8	5.32	4.46	4.07	3.84	3.69	3.58	3.50	3.44	3.39	3.35	3.28	3.22	3.15	3.12	3.08	3.04	3.01	2.97	2.93
9	5.12	4.26	3.86	3.63	3.48	3.37	3.29	3.23	3.18	3.14	3.07	3.01	2.94	2.90	2.86	2.83	2.79	2.75	2.71
10	4.96	4.10	3.71	3.48	3.33	3.22	3.14	3.07	3.02	2.98	2.91	2.85	2.77	2.74	2.70	2.66	2.62	2.58	2.54
11	4.84	3.98	3.59	3.36	3.20	3.09	3.01	2.95	2.90	2.85	2.79	2.72	2.65	2.61	2.57	2.53	2.49	2.45	2.40
12	4.75	3.89	3.49	3.26	3.11	3.00	2.91	2.85	2.80	2.75	2.69	2.62	2.54	2.51	2.47	2.43	2.38	2.34	2.30
13	4.67	3.81	3.41	3.18	3.03	2.92	2.83	2.77	2.71	2.67	2.60	2.53	2.46	2.42	2.38	2.34	2.30	2.25	2.21
14	4.60	3.74	3.34	3.11	2.96	2.85	2.76	2.70	2.65	2.60	2.53	2.46	2.39	2.35	2.31	2.27	2.22	2.18	2.13
15	4.54	3.68	3.29	3.06	2.90	2.79	2.71	2.64	2.59	2.54	2.48	2.40	2.33	2.29	2.25	2.20	2.16	2.11	2.07
16	4.49	3.63	3.24	3.01	2.85	2.74	2.66	2.59	2.54	2.49	2.42	2.35	2.28	2.24	2.19	2.15	2.11	2.06	2.01
17	4.45	3.59	3.20	2.96	2.81	2.70	2.61	2.55	2.49	2.45	2.38	2.31	2.23	2.19	2.15	2.10	2.06	2.01	1.96
18	4.41	3.55	3.16	2.93	2.77	2.66	2.58	2.51	2.46	2.41	2.34	2.27	2.19	2.15	2.11	2.06	2.02	1.97	1.92
19	4.38	3.52	3.13	2.90	2.74	2.63	2.54	2.48	2.42	2.38	2.31	2.23	2.16	2.11	2.07	2.03	1.98	1.93	1.88
20	4.35	3.49	3.10	2.87	2.71	2.60	2.51	2.45	2.39	2.35	2.28	2.20	2.12	2.08	2.04	1.99	1.95	1.90	1.84
21	4.32	3.47	3.07	2.84	2.68	2.57	2.49	2.42	2.37	2.32	2.25	2.18	2.10	2.05	2.01	1.96	1.92	1.87	1.81
22	4.30	3.44	3.05	2.82	2.66	2.55	2.46	2.40	2.34	2.30	2.23	2.15	2.07	2.03	1.98	1.94	1.89	1.84	1.78
23	4.28	3.42	3.03	2.80	2.64	2.53	2.44	2.37	2.32	2.27	2.20	2.13	2.05	2.01	1.96	1.91	1.86	1.81	1.76
24	4.26	3.40	3.01	2.78	2.62	2.51	2.42	2.36	2.30	2.25	2.18	2.11	2.03	1.98	1.94	1.89	1.84	1.79	1.73
25	4.24	3.39	2.99	2.76	2.60	2.49	2.40	2.34	2.28	2.24	2.16	2.09	2.01	1.96	1.92	1.87	1.82	1.77	1.71
26	4.23	3.37	2.98	2.74	2.59	2.47	2.39	2.32	2.27	2.22	2.15	2.07	1.99	1.95	1.90	1.85	1.80	1.75	1.69
27	4.21	3.35	2.96	2.73	2.57	2.46	2.37	2.31	2.25	2.20	2.13	2.06	1.97	1.93	1.88	1.84	1.79	1.73	1.67
28	4.20	3.34	2.95	2.71	2.56	2.45	2.36	2.29	2.24	2.19	2.12	2.04	1.96	1.91	1.87	1.82	1.77	1.71	1.65
29	4.18	3.33	2.93	2.70	2.55	2.43	2.35	2.28	2.22	2.18	2.10	2.03	1.94	1.90	1.85	1.81	1.75	1.70	1.64
30	4.17	3.32	2.92	2.69	2.53	2.42	2.33	2.27	2.21	2.16	2.09	2.01	1.93	1.89	1.84	1.79	1.74	1.68	1.62
40	4.08	3.23	2.84	2.61	2.45	2.34	2.25	2.18	2.12	2.08	2.00	1.92	1.84	1.79	1.74	1.69	1.64	1.58	1.51
60	4.00	3.15	2.76	2.53	2.37	2.25	2.17	2.10	2.04	1.99	1.92	1.84	1.75	1.70	1.65	1.59	1.53	1.47	1.39
120	3.92	3.07	2.68	2.45	2.29	2.17	2.09	2.02	1.96	1.91	1.83	1.75	1.66	1.61	1.55	1.50	1.43	1.35	1.25
∞	3.84	3.00	2.60	2.37	2.21	2.10	2.01	1.94	1.88	1.83	1.75	1.67	1.57	1.52	1.46	1.39	1.32	1.22	1.00

v_1 = degrees of freedom for numerator v_2 = degrees of freedom for denominator

e.g., *Prob.* $(F_{9,16} \geqslant 2.54) = 0.05$.

ν_2 \ ν_1	1	2	3	4	5	6	7	8	9	10	12	15	20	24	30	40	60	120	∞
1	4052	4999·5	5403	5625	5764	5859	5928	5982	6022	6056	6106	6157	6209	6235	6261	6287	6313	6339	6366
2	98·50	99·00	99·17	99·25	99·30	99·33	99·36	99·37	99·39	99·40	99·42	99·43	99·45	99·46	99·47	99·47	99·48	99·49	99·50
3	34·12	30·82	29·46	28·71	28·24	27·91	27·67	27·49	27·35	27·23	27·05	26·87	26·69	26·60	26·50	26·41	26·32	26·22	26·13
4	21·20	18·00	16·69	15·98	15·52	15·21	14·98	14·80	14·66	14·55	14·37	14·20	14·02	13·93	13·84	13·75	13·65	13·56	13·46
5	16·26	13·27	12·06	11·39	10·97	10·67	10·46	10·29	10·16	10·05	9·89	9·72	9·55	9·47	9·38	9·29	9·20	9·11	9·02
6	13·75	10·92	9·78	9·15	8·75	8·47	8·26	8·10	7·98	7·87	7·72	7·56	7·40	7·31	7·23	7·14	7·06	6·97	6·88
7	12·25	9·55	8·45	7·85	7·46	7·19	6·99	6·84	6·72	6·62	6·47	6·31	6·16	6·07	5·99	5·91	5·82	5·74	5·65
8	11·26	8·65	7·59	7·01	6·63	6·37	6·18	6·03	5·91	5·81	5·67	5·52	5·36	5·28	5·20	5·12	5·03	4·95	4·86
9	10·56	8·02	6·99	6·42	6·06	5·80	5·61	5·47	5·35	5·26	5·11	4·96	4·81	4·73	4·65	4·57	4·48	4·40	4·31
10	10·04	7·56	6·55	5·99	5·64	5·39	5·20	5·06	4·94	4·85	4·71	4·56	4·41	4·33	4·25	4·17	4·08	4·00	3·91
11	9·65	7·21	6·22	5·67	5·32	5·07	4·89	4·74	4·63	4·54	4·40	4·25	4·10	4·02	3·94	3·86	3·78	3·69	3·60
12	9·33	6·93	5·95	5·41	5·06	4·82	4·64	4·50	4·39	4·30	4·16	4·01	3·86	3·78	3·70	3·62	3·54	3·45	3·36
13	9·07	6·70	5·74	5·21	4·86	4·62	4·44	4·30	4·19	4·10	3·96	3·82	3·66	3·59	3·51	3·43	3·34	3·25	3·17
14	8·86	6·51	5·56	5·04	4·69	4·46	4·28	4·14	4·03	3·94	3·80	3·66	3·51	3·43	3·35	3·27	3·18	3·09	3·00
15	8·68	6·36	5·42	4·89	4·56	4·32	4·14	4·00	3·89	3·80	3·67	3·52	3·37	3·29	3·21	3·13	3·05	2·96	2·87
16	8·53	6·23	5·29	4·77	4·44	4·20	4·03	3·89	3·78	3·69	3·55	3·41	3·26	3·18	3·10	3·02	2·93	2·84	2·75
17	8·40	6·11	5·18	4·67	4·34	4·10	3·93	3·79	3·68	3·59	3·46	3·31	3·16	3·08	3·00	2·92	2·83	2·75	2·65
18	8·29	6·01	5·09	4·58	4·25	4·01	3·84	3·71	3·60	3·51	3·37	3·23	3·08	3·00	2·92	2·84	2·75	2·66	2·57
19	8·18	5·93	5·01	4·50	4·17	3·94	3·77	3·63	3·52	3·43	3·30	3·15	3·00	2·92	2·84	2·76	2·67	2·58	2·49
20	8·10	5·85	4·94	4·43	4·10	3·87	3·70	3·56	3·46	3·37	3·23	3·09	2·94	2·86	2·78	2·69	2·61	2·52	2·42
21	8·02	5·78	4·87	4·37	4·04	3·81	3·64	3·51	3·40	3·31	3·17	3·03	2·88	2·80	2·72	2·64	2·55	2·46	2·36
22	7·95	5·72	4·82	4·31	3·99	3·76	3·59	3·45	3·35	3·26	3·12	2·98	2·83	2·75	2·67	2·58	2·50	2·40	2·31
23	7·88	5·66	4·76	4·26	3·94	3·71	3·54	3·41	3·30	3·21	3·07	2·93	2·78	2·70	2·62	2·54	2·45	2·35	2·26
24	7·82	5·61	4·72	4·22	3·90	3·67	3·50	3·36	3·26	3·17	3·03	2·89	2·74	2·66	2·58	2·49	2·40	2·31	2·21
25	7·77	5·57	4·68	4·18	3·85	3·63	3·46	3·32	3·22	3·13	2·99	2·85	2·70	2·62	2·54	2·45	2·36	2·27	2·17
26	7·72	5·53	4·64	4·14	3·82	3·59	3·42	3·29	3·18	3·09	2·96	2·81	2·66	2·58	2·50	2·42	2·33	2·23	2·13
27	7·68	5·49	4·60	4·11	3·78	3·56	3·39	3·26	3·15	3·06	2·93	2·78	2·63	2·55	2·47	2·38	2·29	2·20	2·10
28	7·64	5·45	4·57	4·07	3·75	3·53	3·36	3·23	3·12	3·03	2·90	2·75	2·60	2·52	2·44	2·35	2·26	2·17	2·06
29	7·60	5·42	4·54	4·04	3·73	3·50	3·33	3·20	3·09	3·00	2·87	2·73	2·57	2·49	2·41	2·33	2·23	2·14	2·03
30	7·56	5·39	4·51	4·02	3·70	3·47	3·30	3·17	3·07	2·98	2·84	2·70	2·55	2·47	2·39	2·30	2·21	2·11	2·01
40	7·31	5·18	4·31	3·83	3·51	3·29	3·12	2·99	2·89	2·80	2·66	2·52	2·37	2·29	2·20	2·11	2·02	1·92	1·80
60	7·08	4·98	4·13	3·65	3·34	3·12	2·95	2·82	2·72	2·63	2·50	2·35	2·20	2·12	2·03	1·94	1·84	1·73	1·60
120	6·85	4·79	3·95	3·48	3·17	2·96	2·79	2·66	2·56	2·47	2·34	2·19	2·03	1·95	1·86	1·76	1·66	1·53	1·38
∞	6·63	4·61	3·78	3·32	3·02	2·80	2·64	2·51	2·41	2·32	2·18	2·04	1·88	1·79	1·70	1·59	1·47	1·32	1·00

INDEX

A

Acceptance region, 167
Acceptance sampling, 142
Action, courses of, 283
Aggregative price index, 276
    unweighted, 276
    weighted, 278
Alternative hypothesis, 166
Analysis of variance (ANOVA), 195 *et seq.*
    one-way, 206
    table, 210
    two-way, 214
Arithmetic mean, 14
Average absolute deviation, 21
Average of price relatives, 280
    Laspeyres, 280
    Paasche, 281
    unweighted, 280
    weighted, 281

B

Base period, 276
    change of, 281
Bayes decision rule, 284
Bayes Theorem, 289
Bayesian decision theory, 283 *et seq.*
Belief, degree of, 39
Bernoulli distribution, 97
Best fit, line of, 238
Beta distribution, 119
    expected value of, 120
    mode of, 120
    variance of, 120
Between groups sum of squares, 210
Between groups variance, 207
Bias, 138
Binomial distribution, 97
    expected value of, 100
    graph of, 102
    normal approximation to, 132
    Poisson approximation to, 112
    variance, 100

C

Central limit theorem, 145
Central tendency, 13
Change of base, 281
Chi-squared statistic, 198
    continuity correction to, 202
    use in significance tests, 199
    use in goodness-of-fit tests, 223
Chi-squared distribution, 199
    expected value of, 199
    variance of, 199
Class frequency, 2
Cluster sampling, 141
Coding of data, 20
Coefficient of correlation, 250
    distribution of, 252
    partial, 260
    population, 252
    rank, 231
    sample, 250
Coefficient of determination, 248
    in multiple regression, 260
    partial, 260
    population, 252
    sample, 248
Combinations, 62
Complement, of a set, 30
Complementary events, 46
Compound event, 42
    probability of, 42
Compound hypothesis, 170
Conditional probability, 54, 89
    distribution, 90
Confidence interval, 146
Contingency tables, 195 *et seq.*
Continuous probability distribution, 72, 116
Continuous random variable, 69
Convenience sampling, 141
Correlation, 248 *et seq.*
    coefficient of, *see coefficient of correlation*
    direct, 252
    inverse, 252

Cost of uncertainty, 286
Course of action, 283
Covariance, 93
Cumulative frequency, 12
Cyclical factor, 262, 275
Cyclical index, 275
Cyclical residual, 275

D

Data, graphical summary of,
　　1 *et seq.*
Deciles, 15
Decision criteria, 283
　　maximin, 283
　　maximum expected pay-off,
　　　284
Decision theory, 283 *et seq.*
Degree of belief, 39
Degrees of freedom, 152
Density function, 74
　　joint, 91
Deseasonalisation, 273
Determination, coefficient of, *see
　　coefficient of determination*
Difference between population
　　　means, tests for, 154
　　large samples, 154
　　small samples, 156
Direct correlation, 252
Discovery sampling, 142
Discrete probability distribution,
　　70, 97 *et seq.*
Discrete random variable, 68
Dispersion, 14
　　measures of, 20
Distribution, 1 *et seq.*
　　Bernoulli, 97
　　Beta, 119
　　Binomial, 97
　　conditional, 90
　　continuous, 72, 116
　　correlation coefficient, 252
　　discrete, 70
　　exponential, 116
　　F (variance ratio), 208
　　frequency, 1 *et seq.*
　　function, 86
　　geometric, 104
　　joint, 88
　　modes of, 86

　　moments of, 78
　　normal, 121
　　　standard, 123
　　Pareto, 83
　　Poisson, 107
　　sampling, 143
　　skewness of, 86
　　Student's *t*, 152
　　symmetry of, 85
　　uniform, 74

E

Elementary event, 35
　　probability of, 39
Empty set, 28
Errors, Types I and II, 187
　　control of, 191
Estimate, 143
Estimation, 137 *et seq.*
Estimator, 143
Events, 35, 283
　　complementary, 46
　　compound, 42
　　exhaustive, 46
　　mutually exclusive, 46
　　probability of, 41
　　relations between, 46
Expectation, 79
Expected opportunity loss, 285
Expected pay-off, posterior, 291
Expected value, 78
　　of composite random variables,
　　　84
　　of perfect information, 286
　　of sample information, 293
Experiment, random, 34
Exponential distribution, 116
　　cumulative density function,
　　　116
　　expected value, 118
　　probability density function,
　　　116
　　variance, 118
Extrapolation, 240

F

F-distribution, 208
　　expected value, 208
F-statistic, 208
Factorials, 60

Finite population correction, 162, 186
Finite set, 28
Freedom, degrees of, 152
Frequency
    expected, 197
    observed, 196
Frequency distributions, 1 *et seq.*
Frequency polygon, 10
Frequency table, 2

G

Geometric distribution, 104
    expected value, 105
    graph, 106
    variance, 105
Geometric mean, 16
Goodness-of-fit, test for, 223
Group(ed) mean, 18
Group(ed) variance, 24

H

Histogram, 1 *et seq.*
    relative frequency, 7
Hypothesis tests, 165 *et seq.*
    acceptance regions in, 167
    alternative hypothesis in, 166
    compound hypotheses in, 170
    errors in, 187
        control of, 191
    for difference between means
        (dependent samples), 179
    for difference between means
        (independent samples), 175
    for equality of means, 173
    for population means, 167
    for equality of proportions, 201
    general procedure, 165
    null hypothesis in, 165
    rejection regions in, 167
    simple hypothesis in, 170
    small samples, 171

I

Independence, 52
Indices, 276 *et seq.*
    aggregative (unweighted), 276
    aggregative (weighted), 278
    cyclical, 275
    Laspeyres, 278

Paasche, 278
    seasonal, 270
Infinite set, 28
Interpolation, 240
Interquartile range, 20
Intersection, of sets, 30
Inverse correlation, 252

J

Joint distributions, 88
Joint probability, 50
    density function, 91
Judgment sampling, 141

L

Laspeyres index, 278
    average of price relatives, 281
Least squares regression line, 237
Least squares trend line, 268
Line of best fit, 238
Linear regression, simple, 236 *et seq.*
Long-term trend, 262

M

Marginal probability, 51, 89
Matched samples
    sign test for, 228
    Wilcoxon signed rank sum test
        for, 229
Maximin decision criterion, 283
Maximum expected pay-off, 284
Means
    arithmetic, 14
    difference between, tests for,
        154, 156, 173, 175
    equality, tests for, 173
    estimation of, 143
    geometric, 16
    of grouped data, 18
    relation with median and
        mode, 87
Mean absolute deviation, 21
Median, 15
    test for, 221
    test, independent samples, 233
    test for randomness, 220
Modality, 86
Mode, 15
Moments, 78
    significance of third, 87

Moving average, 263
Multiple coefficient of determination, 260
Multiple regression, 258
Multistage sampling, 159
Mutually exclusive events, 46

N

Nature, states of, 283
Non-parametric statistics, 219 *et seq.*
Normal distribution, 121 *et seq.*
  approximation to binomial, 132
  cumulative distribution function, 123
  expected value, 123
  probabilities, 125
  standard, 123
  variance, 123
Normal equations, 238
Null hypothesis, 165

O

Observed frequencies, 196
Ogives, 12
Operating characteristic curve, 187
Opportunity loss, 285
  expected, 285

P

Paasche index, 278
  average of price relatives (unweighted), 280
  average of price relatives (weighted), 281
Pareto distribution, 83
Partial coefficient
  of correlation, 260
  of determination, 260
Pay-off, 283
Percentages of trend, 275
Percentiles, 15
Perfect information, expected value of, 286
Permutations, 59
Poisson distribution, 107
  approximation to binomial, 112

  expected value, 108
  variance, 108
Poisson process, 107
Polynomial regression, 258
Pooled variance, 156
Population, 137
Population coefficient of determination, 252
Population mean, test for, 167
Posterior belief, 287
Posterior expected pay-off, 291
Power curve, 187
Power function, 187
Pre-posterior analysis, 293
Price indices, 276 *et seq.*
Price relatives, simple, 276
Prior belief, 287
Probability, 26 *et seq.*
  density function, 74
    joint, 91
  distribution, 70
    conditional, 90
    continuous, 72, 116
    discrete, 70
    function, 86
  evaluation of, 38
  general properties, 39
  joint, 50
  marginal, 51, 89
  of compound events, 41
  of elementary events, 39
  of events, 41
  relative frequency as approximation to, 38, 76
Proportions
  estimation of, 159
  large samples, 181
  small samples, 183
  test for equality of several, 201

Q

Quantity indices, 282
Quartiles, 15

R

Random effects (in time series), 262
Random experiment, 34
Random sampling, 138
  stratified, 139

Random variables, 68
  continuous, 69
  discrete, 68
  expectation, 79
  expected value, 78
  sums of, 90
Randomness
  median test for, 220
  runs test for, 219
Range, 20
  interquartile, 20
Rank correlation coefficient, 231
Ratio to moving average, 270
Regression, 236 *et seq.*
  least squares, 237
  multiple, 258
  parameters, confidence
    intervals for, 242
  polynomial, 258
  predicted values, confidence
    intervals for, 244
  reduction to linear form, 255
  sampling nature of, 241
  simple linear, 236
  standard error of, 241
  variance of, 241
Rejection region, 167
Relations between events, 46
Relative frequency
  as approximation to
    probability, 38, 76
  histogram, 7
Residuals, cyclical, 275
Residual sum of squares, 215
Runs test of randomness, 219

S

Sample, 137
  coefficient of determination,
    248
  information, expected value of,
    293
  mean, 143
  space, 34
Sampling, 137 *et seq.*
  acceptance, 142
  cluster, 141
  convenience, 141
  discovery, 142
  distribution, 145

  error, 138
  fraction, 163
  judgment, 141
  multistage, 139
  objectives, 141
  schemes, 138 *et seq.*
  simple random, 138
  stratified, 139
  systematic, 141
Scatter diagram, 237
Seasonal factor, 262
Seasonal index, 270
Set, 26
  complement, 30
  element of, 27
  empty, 28
  finite, 28
  infinite, 28
  intersection, 30
  member of, 27
  notation, 27
  union, 30
  universe, 30
Sign test, for matched samples,
  228
Simple hypothesis, 170
Simple price relative, 276
Simple random sampling, 138
Skewness, 86
  significance of third moment,
  87
Standard deviation, 22
  for grouped data, 24
Standard error, of regression, 241
Standard normal distribution, 123
  cumulative distribution
    function, 123
  expected value, 123
  probabilities, 126 *et seq.*
  variance, 123
Standardisation, 127
States of nature, 283
Statistics, 1
  descriptive, 1
  non-parametric, 219 *et seq.*
  summary, 13
Statistical decision theory,
  283 *et seq.*
Student's *t*-distribution, 152
Subset, 30

Sums of squares
    between groups, 210
    residual, 215
    total, 210
    within groups, 210
Symmetry, of distributions, 85
Systematic sampling, 141

T

t-distribution, 152
Time series, 262 *et seq.*
Total probability theorem, 287
Total sum of squares, 210
Trend
    line, least squares, 268
    long-term, 262, 263
    percentage of, 275

U

Uncertainty, cost of, 286
Uniform distribution, 74
Union, of sets, 30
Universe set, 30

V

Variable, random, 68
    continuous, 69
    discrete, 68
Variance, 22, 80
    analysis of, 195 *et seq.*
    between groups, 207
    of regression, 241
    pooled, 156
    ratio, 208
    within groups, 206
Venn diagram, 29

W

Wald-Wolfowitz runs test, 234
Weighted aggregative index, 278
Wilcoxon signed rank sum test, 229
Within groups sum of squares, 210
Within groups variance, 206

Z

z-statistic, 252
z-transformation, 252